Severe and Unusual Weather

Severe and Unusual Weather

Joe R. Eagleman

Second Edition

Trimedia Publishing Company
12008 W. 87th, Suite 117
Lenexa, KS 66215
Ph. 913 599-0505

Preface

An understanding of severe and unusual weather should be a fundamental part of everyone's storehouse of knowledge. Very few places on Earth escape the formation of some type of severe or unusual weather which can have a major impact on our personal property and well-being. This second edition has been developed to provide an understanding of the nature of various forms of severe and unusual weather along with information on dealing with the associated hazards.

This book may be used as an introductory textbook for a course on severe and unusual weather, or as a follow-up course after introductory Meteorology, Geography or Earth Science, since it has been established that a widespread interest exists for learning more about this topic. If it is used for a first course, a companion workbook is recommended, *Weather Concepts and Terminology*, Trimedia Publishing Company, to help students overcome the problem of new terms which are used in meteorology. In order to reach a variety of students, professionals and general readers, the descriptive approach is used. The common bond among interested individuals is a desire to learn more about the many severe and unusual aspects of our natural environment.

This second edition incorporates new information and adds a new chapter dealing with such current topics as acid rain, the greenhouse effect and the ozone hole. These fit in with the unusual weather theme, since human contributions to the atmosphere are altering our weather and making it more unusual. New material has also been added on human response to the weather.

The book is divided into three parts. The first includes frontal cyclones, blizzards, severe thunderstorms, tornadoes, lightning and hailstorms. This section acquaints the reader with the nature of such weather events and places some emphasis on dealing with the hazards associated with each of them. In the second section such topics as hurricanes, floods, drought, unusual weather patterns, and storms, including mountainadoes, are considered. The final section covers weather simulation and management with a discussion of weather modification, laboratory tornadoes, designing houses for withstanding strong winds, using wind and solar energy, altered weather and atmosphere from pollutants, and human response to weather.

Many of the chapters begin with a specific human-weather interaction incident that directly involves the reader in a potential weather hazard or event. While these are not meant to be strictly factual or funny, they are designed to emphasize the importance of a knowledge of unusual or severe weather. Many of them actually happened, including the car and its occupants being dropped on a flat roof by a tornado and the two people left stranded on the side of a cliff by a flood.

As we realize more fully the importance of our weather, we appreciate the need for being better informed about its capricious nature. Such information is important, and sometimes critical, to our survival and prosperity.

Joe R. Eagleman

Contents

Preface   v

Part 1   Cyclones, Thunderstorms, and Tornadoes

Chapter   1   Introductory Overview   3

Chapter   2   The Largest Storm on Earth   10
     Scales of Motion   10
     Jetstream   12
     Air Masses   19
     Fronts   21
     Midlatitude Cyclones   22
     Cyclogenesis   25
     Associated Weather   28
     Summary   30

Chapter   3   Blizzards and Chinooks   32
     Blizzards   32
     Types of Blizzards   33
     Associated Precipitation   38
     Winter Storm Forecasting Terms   42
     Examples of Severe Winter Weather   43
     Survival in Blizzards   45
     Chinooks   50
     Preparation for Winter Storms   52
     Summary   52

Chapter 4 Setting the Stage for Severe Thunderstorms 54
Ordinary and Severe Thunderstorms 55
Atmospheric Stability 56
Moisture Tongue 58
Low Level Jet 59
Upper Air Inversion 61
Midlatitude Cyclone 62
Squall Lines 62
Wind Profile 65
Upper Air Divergence 67
Jetstream 68
Severe Thunderstorm Watches 69
Summary 71

Chapter 5 Nature of Severe Thunderstorms 73
Thunderstorm Development 74
Air Currents within Thunderstorms 75
Thunderstorm Movement 82
Summary 90

Chapter 6 The Strongest Storm on Earth 92
Tornado Development 93
Tornado Damage Paths 96
Pressure and Wind Speed 98
Appearance 99
Distribution and Number 101
Tornado Detection 106
Tornado Safety 111
Summary 112

Chapter 7 Fire from Above 114
Historical Setting 114
Studies by Benjamin Franklin 115
Forms of Lightning 117
Distribution of Thunderstorms 123
Origin of Thunder 123
Development of Electricity in Clouds 124
Nature of the Lightning Flash 126
Lightning Protection 130
Summary 135

Chapter 8 Ice from the Sky 137
Losses from Hail 137
Occurrence of Hailstorms 139
Characteristics of Hailstones 142

Updraft Support of Hailstones   144
Atmospheric Conditions during Hailstorms   146
Hail Producing Thunderstorms   148
Hail Formation   150
Hail Research   151
Summary   159

Part 2   Hurricanes and Other Unusual Weather

Chapter 9   The Mighty Middle-Size Storm   163
Hurricanes, Typhoons and Cyclones   164
Stages of Hurricanes   164
Origin of Hurricanes   166
Some Features of Hurricanes   171
Pressure   172
Wind Speed   173
Size   174
Speed and Lifetime   175
Rainfall   175
Dissipation Over Land   176
Tornadoes in Hurricanes   177
Ocean Swells   179
Summary   180

Chapter 10   Famous Hurricanes and Tropical Cyclones   182
Hurricane Names   182
Exceptional Tropical Storms   183
Hurricane Betsy   184
Hurricane Camille   185
Hurricane Agnes   189
Hurricane Frederic   192
Hurricane Alicia   192
Hurricane Elena   192
Hurricane Hugo   193
Hurricnes Carla   194
Hurricane Donna   194
Hurricane Beulah   196
Galveston Hurricane of 1900   197
Bay of Bengal Cyclone of 1970   200
Summary   202

Chapter 11   Prediction and Modification of Hurricanes   203
Prediction of Hurricanes   203
Paths of Hurricanes   204
Hurricane Path Predictions   206
Remote Sensing of Hurricanes   209
Seeding of Hurricanes   212
Hurricane Safety Rules   216
Summary   217

Chapter 12   Unusual Atmospheric Storms and Synoptic Patterns   219
             Dust Devils   219
             Mountainado   221
             Thermal Vortices   223
             Storm Metamorphosis   224
             Unusual Synoptic Patterns   225
             Heat Waves   228
             Persistent Synoptic Patterns   229
             Unusual Intensity   230
             Unusual Types   231
             Unusual Combinations   234
             Miscellaneous Unusual Weather   235
             Summary   237

Chapter 13   Floods and Drought   239
             Causes of Floods   242
             Repeated Frontal Cyclones   244
             Stationary Front   246
             Individual Thunderstorms   247
             Snowmelt and Other Causes   253
             Flood Warnings   253
             Drought Producing Weather Patterns   256
             Agricultural Drought   259
             Summary   264

Part 3   Weather Simulation, Modification, Management and Human Response

Chapter 14   Urban and Agricultural Weather Modification   269
             Urban Weather Modification   269
             Radiation   270
             Temperature   270
             Relative Humidity   274
             Human and Vegetation Environment   274
             Wind   278
             Composition of the Atmosphere   279
             Precipitation   280
             Summary of Inadvertent Urban Modifications   281
             Intentional Urban Weather Modification   282
             Agricultural Weather Modification   283
             Modification of Water Supply   284
             Modification of Temperature   287
             Modification of Wind and Radiation   289
             Summary   291

Chapter 15 Laboratory Tornadoes   292
            Atmospheric Vortices   292
            Generating Laboratory Vortices   293
            Simulating Winds in Thunderstorms   295
            Laboratory Tornadoes Produced by Blockage   299
            Laboratory Tornadoes from Crosswinds   299
            Variations in Crosswinds   301
            Other Observations of Laboratory Tornadoes   306
            Applications   310
            Summary   312

Chapter 16 Softening the Blow   313
            Topeka Tornado Damage   314
            Lubbock Tornado Damage   323
            Designing Wind Resistant Houses   325
            Summary   331

Chapter 17 Harvesting the Wind and Sunlight   333
            Atmospheric and Other Energy Sources   333
            Wind Power   336
            Solar Energy   342
            Solar Cells   343
            Solar Collectors   344
            Solar Energy Acceptance   347
            Summary   348

Chapter 18 Acid Rain, Ozone Hole and Greenhouse Effect   350
            Occurrence of Acid Rain   350
            Cause and Distribution   351
            Pollution and Precipitation Processes   353
            Politics and Acid Rain   354
            Ozone Hole   355
            The Greenhouse Effect   357
            Concern for Greenhouse Warming   357
            Greenhouse Gasses   358
            Future Atmospheric Temperature   358
            Omitted Factors in the EPA Report   359
            Summary   362

Chapter 19 Human Response to Weather   364
            Blizzards   365
            Tornadoes   366
            Lightning   367
            Hailstorms   367
            Hurricanes   368
            Drought   368
            Floods   369
            Weathering   370
            Temperature   371
            Humidity   371
            Pressure   372

Psycometeorology   373
Mood and Weather   373
Response to Positive Ions   374
Weather Sensitivity   375
Self Test for Weather Sensitivity   376
Conclusions   377

Additional Reading Suggestions   379

Index   391

Part 1
Cyclones, Thunderstorms, and Tornadoes

1
Introductory Overview

We live in a world that is at least occasionally dominated by severe and unusual weather. Many types of severe weather are sufficiently rare that a common defense mechanism of many people is to assume that they will never be directly affected. However, there is hardly a place in the whole world that does not have some peculiar aspect of weather that requires some degree of understanding and preparedness in order to avoid loss of property and, perhaps, even life itself. Fortunately, no particular location has all the different kinds of unusual and severe weather; thus, coastal areas are exposed to the tremendous power of the hurricane that brings high winds and frequently produces flood conditions, while within the interior United States, where hurricanes are not a threat, such severe types of weather as tornadoes, hailstorms, and blizzards are sufficiently frequent that an understanding of these storms is essential when traveling or living in this part of the United States. The complete destruction of a corn crop by hail is shown in Figure 1-1 and houses destroyed by a tornado are shown in Figure 1-2.

Although lightning is a greater hazard in some parts of the world than others, there are very few locations, including Alaska and the Sahara Desert, where occasional severe thunderstorms do not develop numerous lightning strokes. These, of course, are a threat to individuals who may not be properly protected, in addition to the possibility of extensive damage to property and forests, since lightning is known to be responsible for a high percentage of all forest fires. Flash flooding

Figure 1-1. Hailstorms may completely destroy agricultural crops. The field of corn shown above has no chance of recovery from the hailstorm. Hail from a single thunderstorm may affect an area 10 km wide and 30 km long. (Courtesy of J.C. Allen & Son.)

is an increasing hazard as floodplains become more inhabited and urban areas spread into areas prone to flooding (Figure 1-3). In recent years, flash floods have been responsible for an ever increasing number of deaths. On the other hand, drought and prolonged lack of rainfall are common weather related problems in various parts of the world. In many midlatitude locations unusual large scale airflow patterns may cause prolonged drought in areas that ordinarily enjoy adequate amounts of rainfall.

Figure 1-2. Tornadoes are the strongest atmospheric storms and have the ability to destroy whole cities, such as Udall, Kansas, in 1955, or parts of large cities, as shown above for Topeka, Kansas, following the tornado on June 8, 1966. The Washburn University campus in Topeka was almost completely destroyed by the tornado. (Courtesy of the *Topeka Capitol Journal.*)

There are some indications that our weather is becoming more variable with greater drought in the summer months and a return to colder winters. Indeed, some recent measurements and calculations of the average atmospheric temperature in the Northern Hemisphere have shown that the temperature is dropping and has continued to decrease since 1940. This was preceded, however, by an increase in temperature for several decades. This trend indicates that at about 1940, near the middle of the 20th century, we may have enjoyed some of the warmest weather that has occurred over many past centuries.

We are entering an age where we can begin to think of weather simulation, modification, and management, if only on a small scale at

Figure 1-3. Flash floods are a greater threat to human lives and property than is commonly recognized. Flooding may occur so rapidly that individuals seek safety on whatever higher objects are available such as the automobile for the two men shown in this photograph. (Courtesy of Hernando Santos.)

the present time. We can create and control small atmospheric vortices in the laboratory as an aid in understanding the nature of weather phenomena (Figure 1-4). These are useful in studying the behavior of vortices and their relationship to various types of storms in the atmosphere. This may also furnish useful information on the flow characteristics of tornadoes as they destroy houses and other property.

Although we are not able to control atmospheric storms as much as we would like, there are other types of weather that we can definitely label "man-made." These include certain urban and agricultural climates. In urban areas, the climate is modified to a considerable degree by the structures that people have erected. Most of this may be called inadvertent weather modification. Changes include temperature, humidity, and various other characteristics of the environment in urban areas.

Intentional weather modification is currently used on a local scale in many agricultural areas. The most common form of weather modifi-

Figure 1-4. Although weather control is confined to future hopes rather than current reality, it is possible to simulate some of the characteristics of the atmosphere such as that shown here. Laboratory tornadoes can be used to study the specific nature of a vortex, gain information on air currents within a tornado, and perhaps better understand the formation mechanisms of tornadoes.

cation is cloud seeding. Some tests have shown that certain forms of cumulus clouds can be successfully seeded to increase the amount of precipitation, and results of other research have shown that the snow-pack in mountainous areas can be augmented during the wintertime to significantly increase the amount of runoff in the spring.

If we cannot directly manipulate the weather to our liking, it is still possible to protect ourselves in various ways from the damaging effects of severe and unusual weather. This can be done through proper knowledge of the various severe storms, and use of the best information on seeking shelter from them, coupled with appropriate safeguards before the severe weather arrives. The design and construction of houses makes a great deal of difference in their ability to withstand the high winds of hurricanes or tornadoes as well as the heavy load of snow that accumulates on the roofs of many houses during a winter. This is an important protective avenue for those individuals who will someday have their own house designed and constructed.

We may also be able to use the weather to our advantage by harvesting wind and sunlight. Wind power and sunshine represent two important potential sources of available energy; therefore, harvesting the wind and sunlight is an important future part of weather management in the broadest sense.

If you have never stopped to think about the magnitude of the effect of weather on us, you may be surprised to find that in this day of electronic gadgets, we are still quite frequently at the mercy of the elements. In most cases, this need not be the case if we have paid appropriate attention to the nature of the potentially severe weather events in the area where we live or travel, and have informed ourselves about the precautions that are necessary in dealing with this most severe aspect of our environment. Thus, the blizzard can be avoided when traveling during the wintertime by realizing that some weather events must supersede our desires, and travel plans must be modified to include a delay of a day or two to allow a large snowstorm to move through the area. These frontal cyclones are generally forecast quite well by the National Meteorological Center in Suitland, Maryland.

Hurricane warnings are the responsibility of the National Hurricane Center in Miami, Florida. Information gathered there from satellites and reconnaissance flights are used to predict which areas along the coast will be affected by the passing of a hurricane.

Tornado forecasts are the responsibility of the National Severe Storms Forecast Center in Kansas City, Missouri. These storms are much shorter lived than the other two vortex storms previously mentioned, and, therefore, cannot be forecast as individual tornadoes. Areas of the atmosphere that are likely to produce tornadoes, however, can be identified and such forecasts are provided by the National Severe Storms Forecast Center.

It is currently not possible to predict which tropical disturbance will grow into a hurricane or which small thunderstorm will grow into a large tornado-producing thunderstorm. Thus, our knowledge of atmospheric storms and characteristics is incomplete. We should not be discouraged by this, however, but should view this as an opportunity for future studies by ourselves or others. In this book, the nature of various severe and unusual weather phenomena will be described in detail, with emphasis placed on better understanding such weather events before they confront us, in order to provide the information needed to deal with these important, but often neglected, aspects of our environment.

2
The Largest Storm on Earth

SCALES OF ATMOSPHERIC MOTION

Imagine a beautiful snowcovered landscape, and contemplate the many different scales of magnitude that make up the scene. The entire blanket of snow including any large drifts is composed of numerous small individual snowflakes. If we reflect further we know that an even smaller and almost separate world exists that involves the molecular arrangement of the hydrogen and oxygen atoms that form the individual snowflakes. Within the atmosphere different scales of motion, ranging from global air currents to turbulence around a building, frequently affect each other. Evidence is beginning to accumulate that energy is transferred from the larger circulations, such as the jet stream, to smaller circulation systems such as a frontal cyclone. Energy is also transferred from a thunderstorm to a tornado where it is concentrated into a small but fantastic storm.

A physical law that governs the interaction of some of these atmospheric scales of motion is called the Law of Conservation of Angular Momentum (Figure 2-1). The momentum of any moving mass is simply its velocity multiplied by its mass: for a football player his momentum is determined by how fast he is running and his mass. Momentum in an angular sense (for movement in a circular path) is related to the radius of curvature. The Law of Conservation of Angular Momentum tells us that the velocity of a constant mass multiplied by the radius of curvature must always equal a constant value if momentum is conserved. This means that in order to maintain

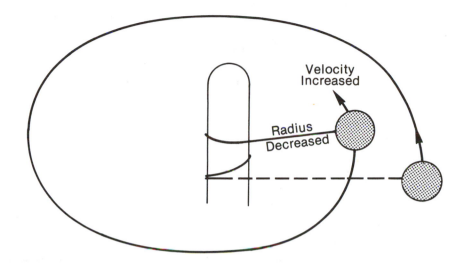

Figure 2-1. The conservation of angular momentum can be illustrated by considering the increase in velocity as a ball on a string wraps around your finger. The velocity of the ball multiplied times the radius must equal a constant if the angular momentum is conserved. If a string attached to a ball is allowed to wrap around your finger, the length of the string, or radius of curvature between your finger and the ball, decreases. The velocity of the ball can be observed to increase in proportion to the reduction in radius of curvature.

constant momentum, any change in the radius must result in a corresponding change in velocity.

You have seen this law applied as ice skaters draw their arms in toward their body to reduce the radius of some of their mass to increase the rate of spin. It can also be demonstrated in a manner more related to atmospheric motion, with a weight on a string; in this case, the radius is the length of the string. If the string is allowed to wrap around your finger as the object on the end of the string rotates, the velocity of the object increases as the string shortens.

Another physical principle (Newton's First Law of Motion) tells us that once an object acquires motion it stays in motion until forces are applied to stop it. These principles continuously operate in the atmosphere because of the various scales of motion involved.

High above the earth's surface are fast-moving streams of air that are constantly interacting with smaller scale weather phenomena in such a way that some of their momentum is transferred down to circulations having a smaller radius. As this occurs, the velocities of the smaller circulations must increase if angular momentum is conserved.

This begins to explain why the very high wind speeds experienced in smaller storms such as tornadoes are possible.

Let us consider the magnitude of some of the different scales of motion in the atmosphere. One of the largest is the jetstream, a portion of which may have a radius of 2,000 km.* A frontal cyclone is a rotating mass of air with a low pressure center that receives some of its momentum from the larger circulation of the jetstream. The circulation radius of the low pressure system is only about half that of the jetstream, or about 1,000 km. Individual thunderstorms grow within a frontal cyclone and have a radius of only about 10 km while a tornado is another order of magnitude smaller with a radius of perhaps 1 km. A few computations will convince you of the possibility of some extremely high velocities as the radii decrease if the momentum remains constant.

Using the radii of curvature just described, a cyclone could contain winds of 200 km/h if the jetstream were 100 km/h. If momentum is conserved and energy is fed to even smaller circulations, their calculated velocities become unrealistically large: 20,000 km/h for the thunderstorm and 200,000 km/h for the tornado. Such velocities are not developed because the transfer of momentum and energy to the smaller circulations is very inefficient in the atmosphere, and the connection between the circulation within a frontal cyclone and a thunderstorm is very weak. The actual mechanics of airflow around thunderstorms are to be discussed in a later chapter, but a more realistic calculation can be made at this point by using a rotational speed of a thunderstorm somewhat less than the speed of the jetstream flowing around it (Figure 2-2). If the radius of rotation of the thunderstorm is 10 km and the circulation speed is only half that of the jetstream, 50 km/h for example, the speed of the tornado winds could be 500 km/h for a large tornado and even higher for smaller tornadoes if momentum were conserved.

JETSTREAM

The larger atmospheric circulations must also have an energy source. This source is the sun, as it exerts its effect on weather systems. The more direct radiation from the sun over equatorial regions is absorbed

---

* Kilometers (km) can be readily converted to miles since 1 km is approximately 0.6 miles.

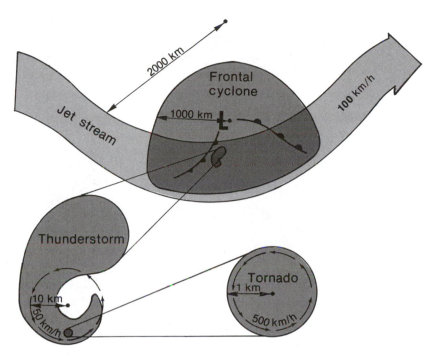

Figure 2-2. Smaller circulation in the atmosphere may acquire greater wind speeds than large circulation because of the conservation of angular momentum. The winds within a tornado of 1 km radius would be 500 km/h if angular momentum were conserved from the larger thunderstorm rotation as shown above. This illustration assumes that momentum was conserved between the thunderstorm and tornado, but not between the jetstream and thunderstorm.

there in large amounts to cause tremendous heating, while polar regions experience low energy levels due to lack of sunshine. This creates a large difference in the amount of heat in different parts of the atmosphere. In fact, recent satellite measurements show excessive heating of the entire area from 32° N latitude to the Equator while net cooling occurs from this latitude to the North Pole (Figure 2-3). Similar differences develop in the Southern Hemisphere.

The amount of energy involved in the unequal absorption of sunlight is tremendous. Heat must be continuously transported northward from equatorial regions or they would continually get hotter, while polar regions would continue to cool. The midlatitudes provide the locus for energy transport; hence the higher incidence of storms and active weather systems in this area. Atmospheric energy

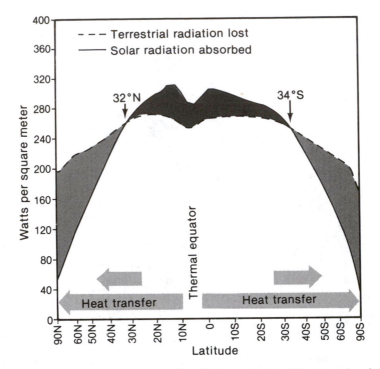

Figure 2-3. Comparison of the amount of radiation received and lost at various latitudes reveals that net cooling occurs poleward from 32° latitude. Heat is transported from the tropics to maintain an equilibrium. These measurements were obtained from the scanning radiometers aboard NOAA Operational Satellite and averaged for the period June 1974 through February 1978. (After NOAA Meteorological Satellite Laboratory, Washington, D.C.)

differences are equalized by heat transported in hot winds moving northward, by cold air moving southward, and through heat associated with water vapor. Energy is absorbed as water evaporates and is released as water condenses to liquid or to the solid form within clouds. This energy (latent heat) amounts to about 2400 J (575 cal) for every gram of raindrops condensed and 2834 J (677 cal) for every gram of snowflakes that form. The amount of heat transferred across midlatitudes averages an astronomical 10^{20} cal per day. The actual transfer of heat is not constant each day as periodic weather disturbances are involved. The transfer of heat is made possible through wind currents since they respond to temperature differences.

Winds in the atmosphere blow because of differences in atmospheric pressure. In tropical regions, with a large heat input, the air expands and becomes less dense; in polar areas where heating is decreased, the air is compressed and very dense. The expanding air in the tropics rises and starts to flow northward over the northern hemisphere and southward over the southern hemisphere. In the northern hemisphere, the flow is deflected to the right of its path of motion because of the Coriolis acceleration that results from the rotation of the earth (Figure 2-4). A stream of air thus develops and flows from west to east over midlatitudes at a height of about 12 km (40,000 feet). This air current is called the jetstream. It is related to the surplus heating in tropical regions and the heat deficit in polar regions. This stream of air serves as a powerful source of energy which is a derivative of the primary source, solar energy.

Detailed measurements of the jetstream show that its location is related to the boundary between the expanded tropical and the denser polar air. If we consider the nature of a vertical slice of air extending from the Equator to the North Pole, the heated air near the Equator is expanded with greater vertical distance between constant pressure levels. The jetstream flows where the horizontal pressure gradient is the greatest and this occurs above midlatitudes near 200

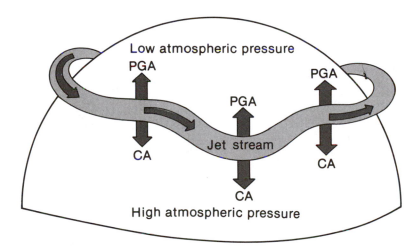

Figure 2-4. The jetstream at a height of 12 km is in a region where the pressure is much lower toward the north with a resulting pressure gradient acceleration (PGA) directed northward. Air would flow in this direction at a height of 12 km if the Earth were not rotating. Due to the Earth's rotation, the Coriolis acceleration (CA) opposes the pressure gradient acceleration and allows the persistence of a westerly air current, the jetstream.

mb (Figure 2-5). The polar jet which separates these two contrasting air masses affects all midlatitude weather systems. In fact, many surface weather systems are generated by the appropriate dynamics of the jetstream which will be described in later sections.

The jetstream is wide enough to reach from Topeka, Kansas, to Oklahoma City, Oklahoma, for example, as shown in Figure 2-6. The jetstream may be considered to have shells of decreasing wind velocity surrounding the inner core where the greatest wind speed is located. Outward from the inner core the wind speed decreases gradually. The velocities shown are high since the greatest recorded jetstream speeds are about 600 km/h. The average speed is only about 65 km/h in the summertime and 130 km/h in the wintertime. Its speed is related to solar heating also; in winter, the North Pole is tilted away from the sun and the contrast in air temperature is much greater, thus strengthening the jetstream.

The position of the jetstream changes with season. During the summer, it moves farther northward, and then comes back southward during the winter. Midlatitude cyclone activity and tornadoes migrate with the season because of this phenomenon. If these storms occur during the winter, they are likely to be located farther south than during the summer.

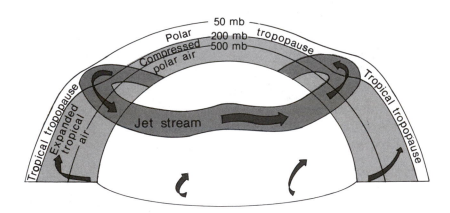

Figure 2-5. The jetstream flows near the 200 mb level because the greatest pressure gradient between the warm expanded air to the south and the cold air to the north exists there. The contrasting air masses help determine the latitude of its location although the jetstream has momentum after its formation and can redistribute large air masses.

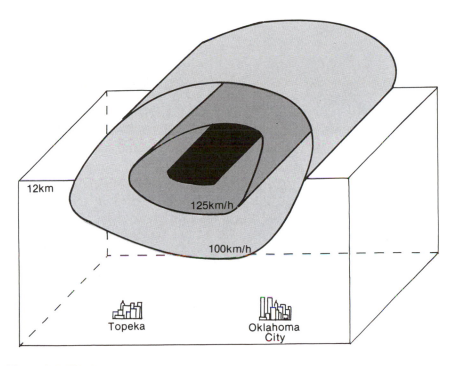

Figure 2-6. The jetstream has a central core that contains wind speeds of perhaps 150 km/h. This core wind speed varies with time and location. Surrounding the central core are winds of lesser speeds. Such bands of strong westerly winds may affect large areas of the United States.

The jetstream does not always travel the same path even during a particular season. It frequently goes through cycles (Figure 2-7) ranging from west-east flow to a meandering pattern that becomes more pronounced with time until the bends eventually are bypassed by the airflow, thus, returning the jetstream to a more west-east flow. It may take the jetstream a month to go through such a cycle or it may become stationary in one of these patterns for the duration of a season. During the summer of 1976, the jetstream developed a persistent ridge (a region of anticyclonic or clockwise curvature) over the United States with related drought conditions. This same pattern occurred in 1965 over New York and other eastern states as a ridge formed over the eastern part of the United States and stayed there for months, with very dry weather as a result.

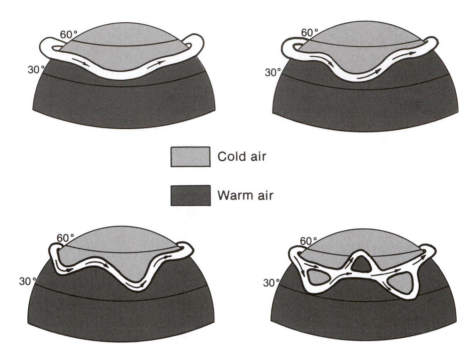

Figure 2-7. The jetstream may complete a cycle consisting of larger meanders until pools of warm and cold air are eventually cut off as the path of the jetstream straightens.

On the other hand, a trough in the jetstream creates very wet weather. The trough is a region of the jetstream with cyclonic or counterclockwise curvature. If a trough becomes located over an area, and is stationary, midlatitude cyclones and repeated cloudy skies with more precipitation are generated. The warm air tends to remain to the south of the jetstream with cold air to the north since this arrangement maintains the strength of the jetstream. As the jetstream goes through cycles, air masses are redistributed as meanders in the jetstream allow cold air to move southward while the warm air moves northward as ridges in the jetstream push northward.

As large meanders develop in the jetstream, unusual weather occurs in various locations. For example, Alaska may be experiencing quite warm weather while the central United States has very cold weather because of large meanders in the jetstream. Very cold weather is typically associated with a large trough in the jetstream that brings extremely cold air southward. The jetstream is related to air masses in a dual fashion since differences in air masses contribute to the development of the jetstream, but after the jetstream gains momentum, with winds of perhaps 200 km/h, it is influential in redistributing heat and air masses, as well as in generating storms. Thus, differences in temperature are required to generate the jetstream, but it also exerts an important influence on the movement of warm and cold air after it develops.

AIR MASSES

Air masses are simply large pools of uniform air, usually occupying several thousand kilometers. Since a particular air mass has similar temperature and humidity throughout, these properties are used to categorize air masses. A continental (c) air mass is very dry while a maritime (m) air mass is quite humid. This humidity characteristic was acquired over the source region, such as over the ocean where considerable evaporation of water increases the humidity of the maritime air or over land where the air gains much less humidity. Temperature characteristics are also acquired in an air mass's source region. A continental air mass can be either tropical (T) or polar (P), depending on its source region that determines its hot or cold temperature. Maritime air masses are also either tropical (hot) or polar (cold). The four major designations of air masses, therefore, are: mT, cT, mP, and cP. Additional air mass types are occasionally important in midlatitudes including continental Arctic (A) from north of the Arctic Circle and equatorial (E) air masses from near the equator. An invasion of Arctic air causes temperatures to fall to very low values in the winter.

The temperature and humidity of the various air masses are quite different in the upper troposphere as well as near the earth's surface. Air from Miami, for example, is quite hot at the surface and decreases in temperature above the surface while continental polar air masses from Canada are much colder at the surface with corresponding colder temperatures in the upper atmosphere.

The major source regions (Figure 2-8) for the United States are the Gulf of Mexico for maritime tropical air masses, from which most of the precipitation east of the Rockies originates; the Desert Southwest for summertime continental tropical air, which is quite warm, with about the same temperature as the Gulf air, but with much less humidity; the Pacific for maritime tropical air, but the mountainous regions limit its influence to mainly along the west coast; the Pacific

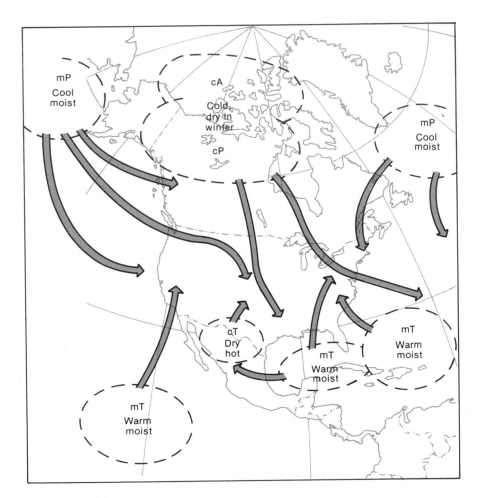

Figure 2-8. Air masses acquire their particular temperature and humidity characteristics from within their source regions as shown here. As they move out of their source regions they affect the weather of large areas.

Ocean for maritime polar air, which sometimes crosses Canada and moves across the United States; Canada for continental polar air masses; and northern Canada and the Arctic for the dryer, and extremely cold, Arctic air that occasionally reaches the United States.

Because of the prevailing wind directions from the west in the upper atmosphere over the United States, the influence of maritime tropical air from the Atlantic does not extend far inland. This air mass source is very important along the East Coast as low pressure systems move northward interacting with different air masses.

FRONTS

As you are well aware, from watching the TV weatherman, fronts generally usher in a different type of weather. Weather fronts represent the boundary between warm and cold air masses; if the warm air is replacing cold air, a warm front exists, while a cold front is the leading edge of a cold air mass as it replaces warm air. The leading edge of a cold air mass takes the form of a thin wedge of cold air as it replaces a warm air mass. As a cold front advances, the trailing wedge of cold air pushes warm air upward where expansion cooling causes clouds to form (Figure 2-9). Since the warm air is advancing and flowing over cooler air as a warm front passes, the type of weather is different from that associated with a cold front. Warm fronts do not force the air to rise as rapidly, but still cause cloud formation. A warm front generates rains that may last all day, followed by warmer temperatures after the passage of the front. Such warm fronts are not as frequent as cold fronts.

The average slope of the leading edge of cold air behind a cold front can be described in terms of its horizontal and vertical dimensions. The slope is about 1 km vertically for every 100 km in horizontal distance. Thus, after a cold front passes and has traveled about

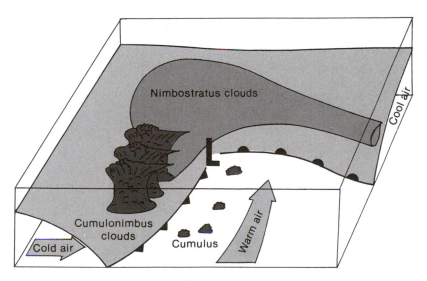

Figure 2-9. Southward moving cold air mass will create a cold front and contribute to thunderstorm formation, while a northward moving air mass will override cooler air to create a warm front and form nimbostratus clouds.

100 km past us, vertical temperature measurements reveal warmer air about 1 km above us. The slope of the boundary between the cold and warm air ahead of a warm front is less and causes a gentle uplift of air flowing over the wedge of cooler air. The slope is about 1 km vertically for every 200 km horizontally. Stratus type clouds are more likely with warm fronts, because of the gentle uplift, while thunderstorms are more likely to develop along cold fronts because of the more rapid uplift of air.

The forward speed of a typical cold front is about 40 km/h; it is slower for warm fronts, about 30 km/h. Either may stall and become a stationary front. In addition to temperature changes accompanying the passage of weather fronts, the wind direction also changes; normally from the southwest to the northwest after a cold front passes and from the southeast to the southwest as a warm front passes. These various types of weather fronts are typically associated with midlatitude cyclones.

MIDLATITUDE CYCLONES

A midlatitude or frontal cyclone consists of an area of low atmospheric pressure that forms the core of the largest type of atmospheric storm on earth. This storm contains lighter winds than smaller vortices but it is a vortex, just as a tornado is a vortex, since it contains horizontally rotating winds around a vertical core. This storm may stretch from the Canadian border to Mexico. A midlatitude cyclone is typically accompanied by a cold front which extends toward the southwest and a warm front which typically extends toward the east, as shown for January 17, 1979, in Figure 2-10. The winds blow from the southwest within the warm air mass located to the south of the low. North of the low pressure center winds are typically from the east, while they blow from the northwest behind the cold front. Fronts are frequently associated with the cyclone in a variety of ways as shown for March 3, 1979, in Figure 2-10.

A midlatitude cyclone does not develop and remain in one location, but moves across the country with a speed and direction that depends on the larger scale circulation of the atmosphere. If the jetstream is strong, the cyclone moves faster, and vice versa. The direction of movement is also related to the jetstream since frontal cyclones travel with the jetstream.

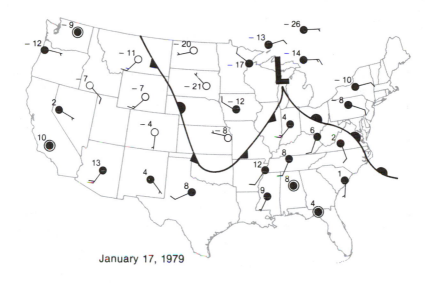

January 17, 1979

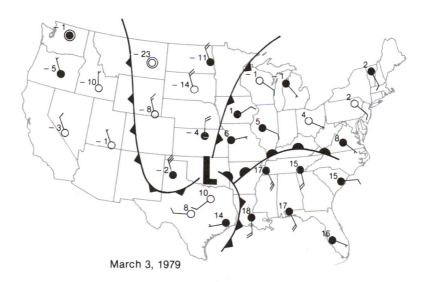

March 3, 1979

Figure 2-10. A typical frontal cyclone (January 17, 1979) has a warm front extending south-eastward and a cold front extending southwestward. Fronts are frequently associated with cyclones in various other ways, however, as shown in the synoptic chart for March 3, 1979.

The cold front normally moves faster than the warm front and eventually catches it. This results in another type of weather front called an occluded front. As the cold front overtakes the warm front, the warm air is lifted upward and frequently continues to cause precipitation as the air swirls around the low pressure.

The center of low pressure is located by plotting the measured atmospheric pressure on a weather map, drawing lines of constant pressure, and thus delineating low pressure centers. The lowest pressure with closed isobars (lines of constant pressure) corresponds to the center of a midlatitude cyclone. Cold fronts may be located by looking for regions where the wind directions shift from the southwest to the northwest. A line separating such wind directions usually represents a cold front although the temperature field must also be inspected. The temperature should be warmer southward from a warm front and colder northward or westward from a cold front.

The movement of a midlatitude cyclone can be forecast in several ways. One method is simply persistence; consider the past speed and direction of the cyclone and then project this trend into the future. Another method consists of using the winds in the upper atmosphere at the 500 mb level. Average surface atmospheric pressure is about 1000 mb (defined as 1013 mb), therefore, the 500 mb level is halfway through the atmosphere and is located at an altitude of about 5.5 km (18,000 feet). The flow of air at this height is similar in direction, but slower in speed than the flow of air at the jetstream. Airflow at 500 mb can be used to forecast the speed of a cyclone located beneath a trough in the jetstream by projecting the cyclone forward at 50% of this speed. For example, if the air is moving at 100 km/h at 5.5 km, then the movement of the low would be projected at 50 km/h, because the surface movement tends to be about half the wind speed at 500 mb.

If we consider the airflow in a midlatitude cyclone, the surface air flows slightly toward the center of lower pressure as it circulates around it cyclonically or counterclockwise. The surface inflow toward the center feeds the large updraft located over the low pressure area. This updraft is caused by an outflow of air (such as diverging air currents) in the upper atmosphere which then results in an updraft beneath it with inflow of air at the surface to feed it. The ascending air within the large updraft area is cooled by expansion into lower pressure at higher altitudes with resulting cloud formation.

A high pressure area (anticyclone) has just the opposite airflow patterns with converging air in the upper atmosphere that results in descending air above the high pressure center and outflow of diverging air at the surface.

CYCLOGENESIS

Midlatitude cyclones are produced by the appropriate conditions with the formation process called cyclogenesis. Recently it has become clear that cyclogenesis is directly related to the jetstream, and air flow patterns above the surface. Most older explanations of cyclogenesis are based on a "polar front" that develops a wave along the front which then produces a warm front and cold front that finally crest, explaining in some way the formation of a low pressure area.

It is now apparent that a number of factors are operating in the upper atmosphere to initiate cyclogenesis; an important one is upper air divergence (Figure 2-11). Diverging air results from air currents with paths that separate farther and farther apart. The extreme case is air currents traveling in completely opposite directions away from each other; normally, this does not occur over very large areas. However, air currents frequently take slightly different paths that result in an outflow of air in those regions where the air currents are separating. Thus, wind currents that are spreading apart produce directional divergence and this results in less air above the earth; hence, lower pressure unless compensating converging air currents are present beneath the upper air divergence.

Speed divergence produces the same result where differences in speed exist. If, for example, a 70 km/h wind is blowing over one location while the wind speed downstream is 80 km/h, outflow of air occurs between these locations if the wind currents are traveling in the same direction. Thus, speed divergence affects those regions located beneath upper air streams where the wind speed increases.

If we consider typical flow patterns from an upper air ridge downwind to a trough, converging air currents are typical because the streamlines corresponding to the paths of air currents are more tightly packed together in the trough than near the ridge due to greater pressure changes over shorter distances outward from a low. Thus, directional convergence occurs from the ridge to the trough, with directional divergence from the trough downwind to the next ridge.

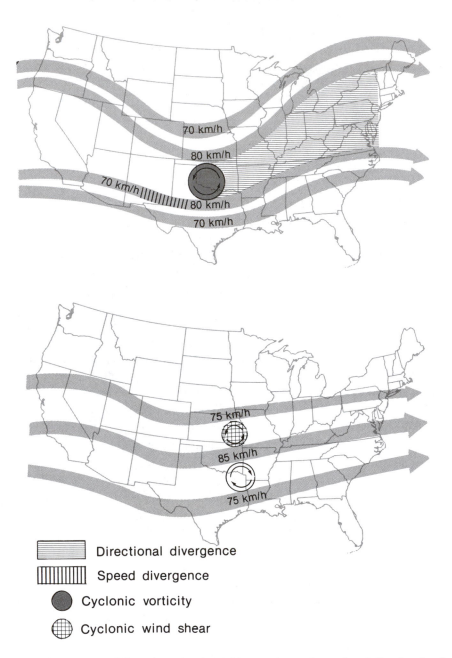

Figure 2-11. Upper airflow dynamics that initiate cyclogenesis consist of directional and speed divergence, cyclonic vorticity, and wind shear. These occur in specific regions of the jetstream as shown here.

Wind speeds are frequently greater in the trough than in the ridge which causes some speed convergence from the trough to the ridge and speed divergence from the ridge to the trough. These opposing factors may counterbalance each other, but more frequently the directional divergence is greater than the speed convergence. Therefore, regions from the trough to an upper air ridge frequently contain divergence that results in lower surface pressure, since less air is located above that point as outflow occurs. A drop in surface pressure produces winds around it because of the pressure difference, and cyclogenesis has occurred.

The jetstream exerts another influence on cyclogenesis by providing cyclonic circulation, also called cyclonic vorticity. Vorticity within the atmosphere can be visualized by considering a large disk placed horizontally in the atmospheric flow patterns (Figure 2-11). If it were located in a trough within a flow field of 80 km/h, for example, it would begin to rotate cyclonically because the curved wind currents would exert more force on the southern side than on the northern side of the disk due to the greater length of contact as shown in Figure 2-11. Such vorticity tends to initiate smaller scale cyclonic circulation. Thus, jetstreams with their large scale cyclonic curvature tend to induce smaller scale cyclonic circulation within the trough of the jetstream.

The jetstream frequently exerts still another influence on cyclogenesis due to wind shear in a horizontal plane which is actually another form of vorticity. If another imaginary disk were placed between two air currents flowing from west to east at 75 km/h on the north side and 85 km/h on the southern side, the stronger winds to the south would cause more rotation than the weaker winds on the north side, thus, producing cyclonic rotation. In this way, cyclonic circulation is initiated north of the core of the jetstream by wind shear, while anticyclonic rotation is generated south of the jetstream by wind shear.

Any one of these three primary factors, upper air divergence, cyclonic vorticity, and cyclonic wind shear, may play the major role in the cyclogenesis of a particular storm or they may contribute in combination. In either case, the generation of midlatitude cyclones is determined by the jetstream. The generation region is typically in the trough of the jetstream, while their life cycle is completed in the region from the trough downwind to the next ridge (Figure 2-12).

ASSOCIATED WEATHER

Frontal cyclones bring a variety of weather with them. Some cyclones are quite weak and produce only cloudy skies with very little precipitation. Others may contain winds that approach 150 km/h (Figure 2-13) and bring fog, rain, sleet, and snow. Severe thunderstorms de-

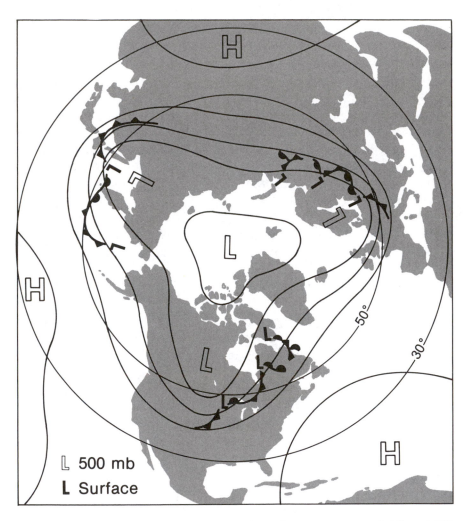

Figure 2-12. Schematic circumpolar chart showing a three wave pattern of airflow at 500 mb with associated frontal cyclones. Such cyclones may develop in the trough and complete their life cycle while traveling toward the next ridge or they may develop beneath the 500 mb low pressure systems.

Figure 2-13. Frontal cyclones may generate winds strong enough to wreck ships, as shown here. Large waves whipped by high wintry winds broke the SS ARGO Merchant in half on December 21, 1976. The 200 m, 18,700 ton tanker was bound for Salem, Massachusetts, with a cargo of 7.3 million gallons of oil when she ran aground 28 km southeast of Nantucket Island in international waters causing a major oil spill. (Courtesy of U.S. Coast Guard.)

veloped within them may also produce lightning, hail, and tornadoes. Several of these associated storms will be the topic of other chapters, including blizzards to be considered in the next chapter.

Frontal cyclones influence large areas as shown in Figure 2-14. The comma shape of the cyclone is apparent from satellite photographs; the head of the comma-shaped cloud pattern corresponds to the center of low pressure, whereas the tail corresponds to the cold front. The jetstream flows over the low pressure center or slightly south of it. Since frontal cyclones are generated near the trough of the jetstream, the earth's surface beneath the region from the trough to the ridge is characterized by wetter than normal weather. In contrast, locations beneath the region from the ridge to the trough will be exposed to drier than normal weather. This information is used by the National Meteorological Center for making 30-day weather outlooks, since the longwave flow patterns of the jetstream frequently remain basically unchanged for several days or weeks.

Figure 2-14. Satellite photographs are helpful in determining the geographical extent and characteristics of frontal cyclones. They frequently have a comma shape with the head of the comma corresponding to the center of low pressure. The low pressure center is located over the Oklahoma panhandle in the above photograph taken on April 20, 1976. The tail of the comma extending southward across eastern Texas has a very distinct western edge indicating the formation of thunderstorms along the cold front. The jetstream frequently blows just south of the center of low pressure. The strong upper level winds help form the cloud streaks over Wisconsin and the Great Lakes. (NOAA photograph.)

If a frontal cyclone is on a path that will take it just north of your location, you will experience first the southeasterly winds prior to passage of the warm front. As the warm front approaches, steady rain is likely. The rain is followed by southwesterly winds and a warmer temperature. As the cold front approaches, thunderstorms develop with brief intense downpours. After the showers pass, the air is much colder with winds from the northwest.

SUMMARY

Many different scales of motion exist in the atmosphere. The jet-stream has a large radius of curvature as it meanders around the earth high above the surface. It is related to differential heating of the earth by the sun, and changes location and speed with the season because of this. The conservation of angular momentum specifies that atmospheric circulations of smaller radii will have greater speeds if momentum is conserved.

Air masses having different temperature and humidity exist at different latitudes and play a significant role in weather events. Cold fronts lift air more rapidly than warm fronts and increase the chances for thunderstorms. Frontal cyclones have fronts associated with them in a variety of ways.

Cyclogenesis occurs in response to such upper air dynamics as horizontal divergence, cyclonic vorticity due to curvature of the jetstream, and cyclonic wind shear. Since these are typically associated with the region just north of a trough in the jetstream, this is the usual location of cyclogenesis.

Frontal cyclones are carried by the upper air currents. Therefore, their usual path is from the southwest toward the northeast since they are generated near cyclonic bends in the jetstream. Frontal cyclones influence large areas of the earth and frequently have the appearance of a large comma when photographed from space.

3
Blizzards and Chinooks

Although the sky was very hazy below the winter clouds, the highway was dry and the road was visible for some distance ahead. Then snow-flakes began to meet the car head on, very slowly at first. But the road ahead soon vanished behind a steady stream of white dots as they dominated the air and then the ground as well. So we will be a few minutes late, we can explain that the weather slowed us down. As the steady stream of snowflakes seemed endless, gusts of wind began to sway the car. The seriousness of the situation struck home only as the car plowed into a snowdrift, deadening the motor from snow sucked into the carburetor. As we surveyed the scene, it was soon apparent that it would be impossible to dig the car out of the snow, even if we were able to get the motor going again.

We had passed some houses a mile or so back; perhaps we could get to one of them to call our motor club for a tow truck. The first few steps were easy, but then the strong northeast wind made its presence known as it bombarded us with snowflakes and made our medium weight clothing feel very thin. As our hands and feet began to feel numb, we realized how vulnerable we had become, and began to desperately search for a way out. . . .

BLIZZARDS

Blizzards frequently accompany severe frontal cyclones in winter, and are generally underestimated when considering severe storms, particularly those responsible for a large amount of damage and numerous deaths in the United States. Many people are surprised to learn that the number of deaths from blizzards (more than 100 per year) is similar to the number of people killed by tornadoes. A single blizzard from January 26 to 28, 1978, killed almost 100 people in the eastern United States. This blizzard was the result of a very intense low pressure system with a central surface pressure of 958 mb and winds up to 134 km/h (89 mi/h). Most of the deaths from winter storms (as well as the additional thousands of deaths each year from overexposure to ordinary cold weather) could be prevented by proper planning during winter travel and by taking other precautions.

A blizzard warning is issued when a frontal cyclone is expected with winds of at least 60 km/h (35 mi/h) and temperatures of less than $-6°C$ ($20°F$) with accompanying snow. A severe blizzard warning is issued if the winds are expected to be at least 75 km/h (45 mi/h) and temperatures are expected to drop below $-12°C$ ($10°F$) with snow.

TYPES OF BLIZZARDS

The typical blizzard cyclone is of the longwave cyclone type, where the surface low pressure system is connected with the low pressure center in the upper troposphere (Figure 3-1). The longwave cyclone develops as the jetstream flows cyclonically around the center of lower pressure in the upper atmosphere with sufficient intensity to extend this upper tropospheric vortex down to the earth's surface, thus creating very intense low pressure there. The resulting surface winds have a much greater speed as they surround the unusually low surface pressure. In addition, they are directly related to the winds of the upper troposphere since a single vortex of circulating air surrounds the common low pressure core that extends from the surface past the jet stream. Such longwave cyclones frequently move across the United States during the spring months, particularly March, causing much higher winds than otherwise occur with blizzard conditions in the northern United States. These storms also develop during other times of the year, and produce heavy snowfall during the coldest winter months, with high winds and cold temperatures creating blizzard conditions.

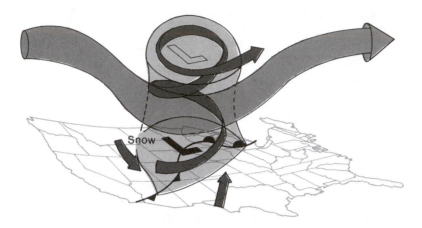

Figure 3-1. Longwave cyclones are frequently blizzard producers, since the surface winds are strong and they travel very slowly. The surface low is connected vertically with the upper air low to form the largest of all atmospheric storms.

The longwave cyclone moves with a very slow forward speed because its rate of forward travel is connected directly to the rate of progressive eastward movement of the upper tropospheric longwave. Since this ordinarily is slower than the movement of other storm systems, the longwave cyclone affects a particular location for a longer period of time, and this, combined with its lower surface pressure, greater pressure gradient with stronger winds, and heavy snow, creates blizzard conditions.

The longwave cyclone shown in Figure 3-2 for March 2 and 4, 1977, traveled a distance of 1600 km in the 48 hour period, an average speed of 33 km/h. Although this low pressure center was not particularly strong, only 997 mb, the winds around it were strong enough to produce blizzard conditions in South Dakota as it passed to the south on March 3rd.

An example of a very intense longwave cyclone is the storm of February 23 and 24, 1977 (Figure 3-3). On the 23rd, the low pressure center was in northeast Kansas with winds of 40 km/h (25 mi/h) common around the center; these strong winds picked up considerable dust from the dry soil in eastern Colorado, New Mexico, Oklahoma, and Texas. The dust picked up by the high winds south of this low pressure system was visible on satellite photographs and could be traced all the way across the southern United States and into the

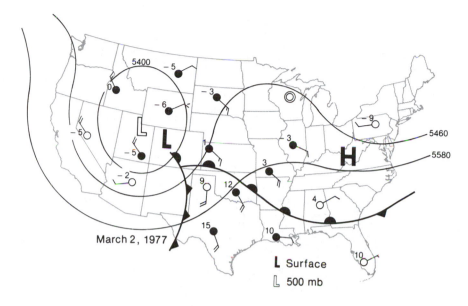

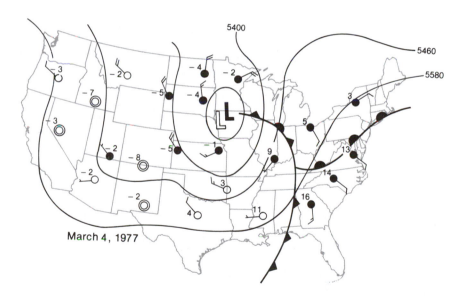

Figure 3-2. This weak, longwave cyclone traveled at about 33 km/h and brought strong winds and snow to the north central United States.

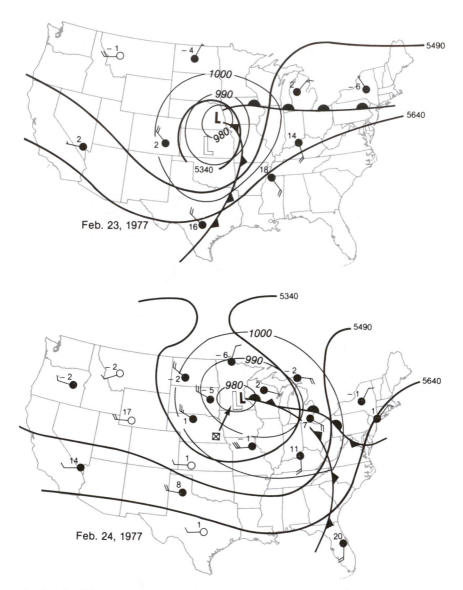

Figure 3-3. This strong longwave cyclone brought blizzard conditions to the north central United States. Winds were also so strong south of this cyclone that great quantities of dust were picked up from southwestern United States and carried across the southern United States into the Atlantic Ocean.

Atlantic Ocean. In some locations, the visibility was reduced to the point that darkness occurred in midmorning. This midlatitude cyclone had a central pressure of 976 mb with a heavy band of snow north of the center of lower pressure. This is common for blizzards and occurs as the warm air south of the low pressure system is carried counterclockwise around the low pressure, where it cools as it moves north of the center of low pressure, and the moisture condenses as snow. Additional precipitation in the form of rain may fall along the warm front, as well as along the southern parts of the cold front.

Another type of midlatitude cyclone, the trough cyclone, also causes blizzard conditions, especially along the east coast. The trough cyclone (Figure 3-4) develops because of all of the upper atmospheric support features described in the previous chapter, including cyclonic vorticity, cyclonic wind shear north of the jetstream axis, and horizontal divergence of air currents in the upper troposphere. These various upper tropospheric factors contribute to an outflow of air aloft, causing a lower pressure at the surface, due to less mass of air above, and this results in air circulation around the lower pressure. These formation factors are most common in the trough of the jetstream; hence the name trough cyclone.

Trough cyclones travel at a higher speed than longwave cyclones because they are carried along by the winds in the upper troposphere.

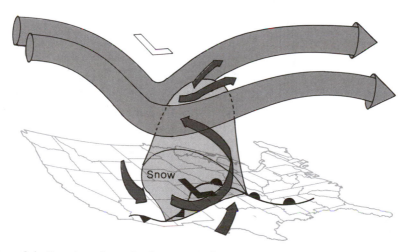

Figure 3-4. Trough cyclones develop as a result of cyclonic vorticity, cyclonic wind shear, and horizontal divergence near the jetstream. They are carried along by the upper level winds and dissipate as they move under the influence of an upper level ridge.

In fact, their forward speed of travel can be roughly estimated by dividing the speed of the 500 mb winds by one-half. A typical speed for trough cyclones is 50 km/h (30 mi/h).

The path of a trough cyclone is also affected by the upper airstream. Trough cyclones cause blizzard conditions all along the east coast as they follow the jetstream. When a jetstream trough develops over the Gulf Coast states, it produces strong south-southwesterly flow along the east coast. This causes cyclones that develop in the trough to travel northeastward following the eastern coastline. Cyclonic surface winds around the low pressure bring air from the Atlantic over the land, providing a strong moisture source that produces heavy rain and snowfall from these storms. Such "northeastern" storms are responsible for blizzard conditions over extended areas of the heavily populated east coast.

Trough cyclones developed at the surface in two separate locations beneath the trough of the jetstream on November 27, 1978 (Figure 3-5). Their different locations beneath the upper airstreams caused their paths to merge producing a single trough cyclone two days later. This cyclone was carried along by the jetstream for another 24 hours until it reached the next ridge where it dissipated.

ASSOCIATED PRECIPITATION

Blizzards not only produce snow, but various other types of precipitation as well. The area of greatest snowfall is usually centered about 200 km north of the low pressure center as moist air from the south travels cyclonically around the low pressure center (Figure 3-6). Southward from the area of heaviest snowfall, sleet and freezing rain commonly form, while still farther southward, rain may fall. Freezing rain on the ground is possible only when the air temperature is above freezing in the cloud where the water is condensing since it must initially be in the liquid form. With surface temperatures below freezing the liquid rain may quickly freeze as it strikes a cold surface. These temperature characteristics are usually associated with an inversion that allows the air near the surface to be cold while the air temperature is warmer at a higher level in the atmosphere. The average annual distributions of snow and freezing rain are shown in Figure 3-7.

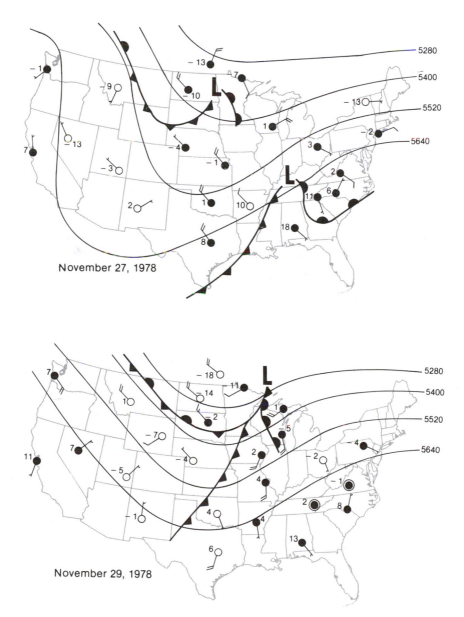

November 27, 1978

November 29, 1978

Figure 3-5. Trough cyclones that developed on November 27, 1978, were guided by the upper air stream that caused them to merge into a single, stronger cyclone two days later.

Ice pellets (sleet) ordinarily develop from an upper atmospheric inversion. With such a temperature profile a warm layer exists, where liquid raindrops can form, above a colder layer. As rain falls down through the cold layer raindrops are frozen into ice, thus producing sleet at the surface. These appropriate temperature profiles contain-

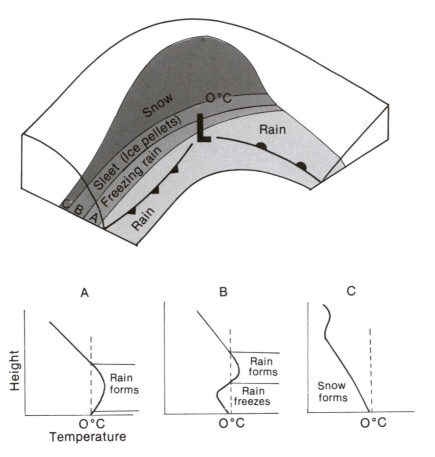

Figure 3-6. Various types of precipitation are associated with a frontal cyclone. The temperature change with height helps determine the type of precipitation at the surface. Ice pellets and freezing rain require warm air at some distance above the surface in order for the water vapor to condense in a liquid form. If the layer of air below the warm air is thick enough and cold enough to allow the rain to freeze, it becomes sleet at the surface; if not it may fall to the ground as rain, or if the surface is at or below the freezing point, it may freeze as it strikes the ground. The formation of snow flakes requires a temperature below freezing for their formation.

Mean annual total snowfall (meters)

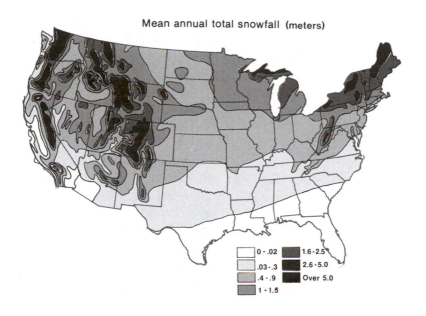

0 - .02	1.6 - 2.5
.03 - .3	2.6 - 5.0
.4 - .9	Over 5.0
1 - 1.5	

Mean annual number of days with glaze (freezing rain)

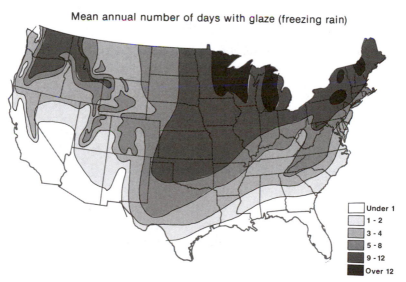

Under 1
1 - 2
3 - 4
5 - 8
9 - 12
Over 12

Figure 3-7. Snowfall is greatest in the mountainous areas of the western United States as orographic effects produce considerable precipitation during the winter months. The extreme southern United States gets little or no snowfall with snowfall amounts increasing northward across the central and eastern United States. The number of days with freezing rain varies from none in the extreme south and southeastern United States to an average of 12 days per year for areas around the Great Lakes and northeastern United States.

ing surface inversions and upper-air inversions occur at varying distances behind the cold front. Surface inversions are most likely near the cold front, while upper-air inversions are likely to develop 200 km or so behind the cold front.

A midlatitude cyclone appears as a comma-shaped storm on a satellite photograph as previously described, with the cloudiness and precipitation near the center of the low pressure forming the head of the comma and the cold front producing the tail as condensation and precipitation form along the frontal system. The type of precipitation associated with a midlatitude cyclone is, of course, related to the temperature of the air. The freezing line frequently passes through the low pressure center with the warm air south of the center and cold air to the north. This structure may change as the storm progresses and the cold air from the north flows cyclonically around the center of low pressure. Thus, blizzard conditions are most common north or west of a low pressure center, while heavy rains fall south of the center.

WINTER STORM FORECASTING TERMS

The National Weather Service uses various terms with specific meanings in order to warn the public of the dangers associated with a winter storm. Ice storm, freezing rain, or freezing drizzle are terms used to alert the public to the possibility of a coating of ice expected to form on the ground, schools, buildings and other objects with the possibility of heavy damage. If the qualifying term, heavy, is used with these terms, it indicates the ice coating will be greater and more intense damage is likely to trees, wires, and other vulnerable objects.

If snow is forecast without a qualifying word, it means that a steady fall of snow is expected to continue for several hours. Heavy snow warnings are issued when a substantial accumulation is expected. This is generally 10 cm (4 inches) of snow or more within the next 12-hour period. Snow flurries are forecast when snow is expected to fall for short durations or intermittent periods. Snow flurries would not be expected to produce a very large accumulation. Snow squalls are brief, intense snow falls that may last for only a short period of time. They may, however, be accompanied by gusty surface winds. Blowing and drifting snow is forecast when strong winds accompany snow or occur when loose snow is on the ground. These can reduce visibility and produce drifts that may be a problem for travel. In the

Northern Plains, the combination of blowing and drifting snow after the snow has ended may be referred to as a ground blizzard.

As previously described, blizzard warnings are issued when the wind speeds are expected to be at least 60 km/h with falling snow and temperatures –6°C or lower. Cold wave warnings are issued when the temperatures are expected to fall rapidly during the next 24 hours. These are issued primarily during the fall months as a warning to provide protection for agricultural crops and other commercial activities.

EXAMPLES OF SEVERE WINTER WEATHER

Severe blizzards occurred during January of 1978 as intense low pressure systems traveled across the United States. One of these, referred to at the beginning of this chapter, was the very strong storm of January 26. This severe blizzard was produced by a longwave cyclone located over the eastern United States (Figure 3-8). The very low central pressure produced a large pressure gradient, thereby surrounding itself with the resulting high winds. The strength of the storm pulled moist air from over the Atlantic to furnish the moisture for heavy snow. The result was snarled traffic, closed businesses and schools, injuries, and loss of lives.

Several severe blizzards also made their presence known during January and February of 1977, producing one of the coldest winters on record. Eastern states such as New York were particularly hard hit by blizzard conditions, with snow accumulated to the second story level of some buildings, chaos created in travel within the city, and many other undesirable effects produced. In fact, thousands of people were unemployed during the winter as industries were closed due to shortages of gas and other fuels resulting from this severe winter. January of 1979 also produced record breaking low winter temperatures. These winters were the coldest of this century over much of the central and eastern two-thirds of the nation.

Mountain blizzards are common as midlatitude cyclones and upper-air troughs move over mountain locations. These come from the northwest, west, or southwest to affect the Rocky Mountain states. On January 22, 1969, many parts of Montana received 45 cm of snowfall as a trough cyclone moved over the mountains.

In December, 1975, many eastern states were hit by a severe blizzard which developed very suddenly on the 19th. Prior to its development, a trough existed in the upper atmospheric flow pattern as

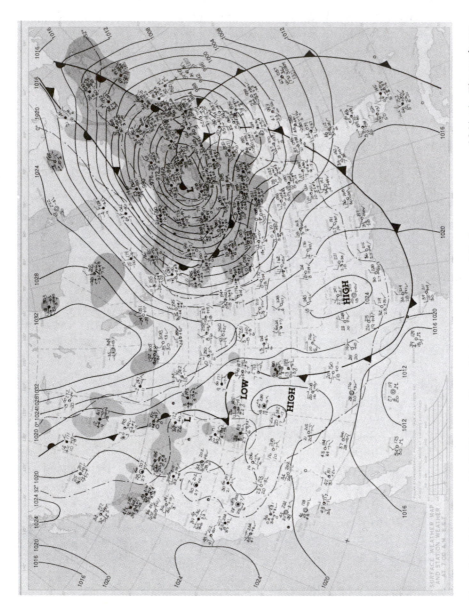

Figure 3-8. This January 26, 1978, blizzard brought strong winds around the low pressure core of 964 mb.. The result was stalled traffic, closed schools and business, and scores of deaths.

the only indication of potential blizzard development, since surface high pressure was located over the Great Lakes area, just north of the upper tropospheric trough (Figure 3-9). By the 20th of December, however, a well developed low surface pressure had replaced the high pressure over the Great Lakes as the upper tropospheric circulation centrifuged the air outward and extended down to the surface, creating a longwave cyclone with blizzard conditions. This blizzard dumped 45 cm of snow on Boston, 35 cm on Albany, and 30 cm on Rochester, in a very short period of time. On the 26th of December, as the storm moved northeastward, it was responsible for 5 cm of glaze throughout much of New Hampshire.

Stationary fronts are frequently thought of in connection with flooding, but they can also produce heavy ice. This occurred on March 4, 1976, throughout the United States in a region extending from Colorado to New York. A trough in the upper troposphere centered over Arizona produced southwesterly winds extending from Texas through Pennsylvania (Figure 3-10). In typical fashion, the stationary front also extended from Texas through Pennsylvania since it was located parallel to the jetstream. The front remained stationary because no forces were acting to move it, setting up the situation for continued icy conditions. Precipitation occurred as southerly winds south of the stationary front brought warm moist air over the frontal surface to be opposed by northerly surface winds. This caused lifting of the moist air, condensation, and severe problems from thunderstorms, icing, and snowfall.

SURVIVAL IN BLIZZARDS

Hypothermia is the name given to the rapid loss of body heat. A person may survive for days, or even weeks, without food and with very little water, but he cannot survive for very long without body heat. Heat loss is also dependent on other factors in addition to air temperatures. The combination of high winds and cold temperatures can cause very rapid loss of heat. It is possible for a person with wet clothing to freeze to death at temperatures of 10°C (50°F). Water conducts heat away very rapidly, and provides additional cooling as latent heat is extracted from skin and clothing for evaporation.

Sipple's equation relates the temperature and wind speed to the cooling power of air. It is most commonly used in the form of the

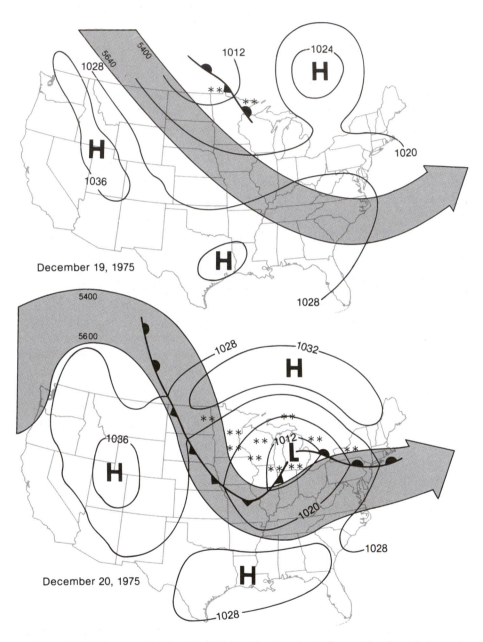

Figure 3-9. Unforecasted blizzard conditions developed rapidly over the Great Lakes and New England on December 20, 1975, as a surface low pressure system replaced a high pressure area located north of the trough of the jetstream.

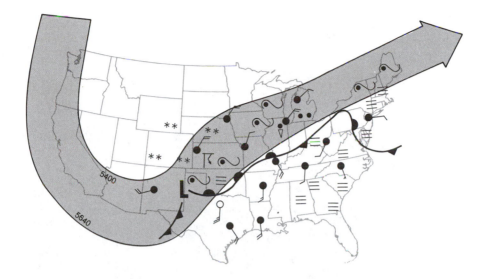

Figure 3-10. Freezing rain fell on March 4, 1976, over an area stretching from Texas to Maine as a front stalled in a parallel position to the jetstream. Continued south winds southward from the front produced a prolonged ice storm.

wind chill factor, which accounts for the influence of winds of varying speed to give the temperature of calm air of comparable cooling power (Figure 3-11). Thus, a temperature of –1°C (30°F) and a wind velocity of 32 km/h (20 mi/h) has the same effect as a temperature of –7°C (20°F) and a wind velocity of only 16 km/h (10 mi/h). Exposed legs, face, or hands will freeze at temperatures of –14°C (–10°F). If the temperature is below –35°C (–30°F) considerable danger of frostbite exists, and this danger increases as the wind speed increases. Such low temperatures will cause flesh covered by ordinary clothing to freeze in only 60 seconds if the wind speeds are very high. A person exposed to very cold temperatures may experience painful tingling, then numbing of the limbs, as their temperature drops below normal body temperature. With continued cooling this will be followed by hallucinations, drowsiness, and death.

Each year thousands of elderly Americans, along with motorists, hikers, and other outdoor enthusiasts, succumb to overexposure to the cold. The chances for accidental hypothermia are greatly increased by several other factors, such as inadequate clothing, inactivity, loss of sleep, little body fat, and the consumption of alcohol, barbiturates, and tranquilizers.

Dry bulb temperature (°C)	Wind Speed (km/h)										
	6	10	20	30	40	50	60	70	80	90	100
	Equivalent temperature (wind chill)										
20	20	18	16	14	13	13	12	12	12	12	12
16	16	14	11	9	7	7	6	6	5	5	5
12	12	9	5	3	1	0	0	-1	-1	-1	-1
8	8	5	0	-3	-5	-6	-7	-7	-8	-8	-8
4	4	0	-5	-8	-11	-12	-13	-14	-14	-14	-14
0	0	-4	-10	-14	-17	-18	-19	-20	-21	-21	-21
-4	-4	-8	-15	-20	-23	-25	-26	-27	-27	-27	-27
-8	-8	-13	-21	-25	-29	-31	-32	-33	-34	-34	-34
-12	-12	-17	-26	-31	-35	-37	-39	-40	-40	-40	-40
-16	-16	-22	-31	-37	-41	-43	-45	-46	-47	-47	-47
-20	-20	-26	-36	-43	-47	-49	-51	-52	-53	-53	-53
-24	-24	-31	-42	-48	-53	-56	-58	-59	-60	-60	-60
-28	-28	-35	-47	-54	-59	-62	-64	-65	-66	-66	-66
-32	-32	-40	-52	-60	-65	-68	-70	-72	-73	-73	-73
-36	-36	-44	-57	-65	-71	-74	-77	-78	-79	-79	-79
-40	-40	-49	-63	-71	-77	-80	-83	-85	-86	-86	-86
-44	-44	-53	-68	-77	-83	-87	-89	-91	-92	-92	-92
-48	-48	-58	-73	-82	-89	-93	-96	-98	-99	-99	-99
-52	-52	-62	-78	-88	-95	-99	-102	-104	-105	-105	-105
-56	-56	-67	-84	-94	-101	-105	-109	-111	-112	-112	-112
-60	-60	-71	-89	-99	-107	-112	-115	-117	-118	-118	-118

Danger from freezing of exposed flesh

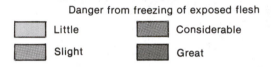

Little   Considerable

Slight   Great

Figure 3-11. The air temperature and wind speed are combined to obtain the wind chill factor or equivalent temperature. An air temperature of 0°C has the same cooling power as a temperature of 8°C if the wind speed is 20 km/h. The danger from freezing of exposed flesh increases as the wind chill increases.

A person's limbs may be able to withstand a drop in temperature of 16°C (30°F), but if his internal body temperature drops more than about 5°C (10°F), death will result. Body heat is conserved as surface blood vessels are constricted by normal response mechanisms. The increased danger of hypothermia with the use of alcohol or certain other drugs results from their opposite action on blood vessels by expansion and increased loss of heat.

Frostbite occurs as the fluid between body cells freezes and the cells become dehydrated. This process is accompanied by feelings of extreme cold, burning sensations, and then numbing. For centuries the suggested treatment has been rubbing the frostbitten area with snow or ice. Very recent tests have shown, however, that it is much better to rewarm the frozen tissue rapidly, in warm water or by another warm part of the body. Frozen tissue should not be rubbed as it can be damaged easily. Overheating a frozen part should, likewise, be avoided.

Prevention of frostbite is far better than suffering the potential consequences that include stunted bone growth or amputation of a limb, besides the possibility of death from hypothermia. Several layers of clothing are more efficient in sealing the body from heat loss than single heavy garments. While gloves are better for driving and other activities, mittens are much better for protecting fingers from frostbite while outdoors. Several pairs of socks are also helpful in protecting the feet from frostbite.

Traveling in severe storms can be extremely serious business (Figure 3-12). It is advisable to plan winter trips accordingly, realizing that blizzards are one of the greatest potential killers of people, as well as livestock and wildlife. Rational winter travel demands that the trip be planned in advance with possible alternate routes. You

Figure 3-12. Driving can become hazardous or impossible during winter snow storms. It is sometimes better to delay travel plans than to risk the consequences of travel during blizzard conditions.

should estimate the time it will take for the trip and give information to someone else concerning your expected arrival time. It is very important to consult weather information and to have a vehicle that is properly equipped. If a blizzard is forecast, with traveler's warnings issued, and a trip is absolutely necessary, then it is advisable to take someone else along or travel with another vehicle if possible.

Driving on snow or ice requires different techniques, such as starting, steering, and braking very gently. You should get the feel of the road and allow plenty of room for stopping. A gentle pumping of the brakes is usually more effective than hard braking that may cause locking of the wheels with skidding. Good visibility is important with winter driving; windshields and rear windows should be clear so that your visibility is not reduced to a peephole through the glass. Headlights should be used while traveling rather than parking lights. Snow tires or chains are helpful in driving on snow or ice. A point to remember is that ice is much slicker near the freezing temperature than with much lower temperatures. A braking distance of twice as long is required on glazed ice at $-1°C$ ($30°F$) than at $-18°C$ ($0°F$).

If you become stuck in snow, do not panic, and beware of overexertion through such efforts as rapid shoveling or trying to manually push the car out of a snowdrift. It is usually best to stay in the vehicle unless you are very close to help that is visible from your automobile. Disorientation and extreme fatigue can occur quickly in blowing and drifting snow; with very cold temperatures this can be disastrous. Remember that carbon monoxide poisoning can be fatal; therefore, it is advisable to allow fresh air into a car at intervals while running the motor and heater sparingly. This is also necessary because a full tank of gas will last only a few hours if run continuously. It is advisable to leave the dome lights or emergency lights on so that the car can be seen by others that may be approaching on the road.

CHINOOKS

If you become completely stranded in a blizzard, you can hope for rescue or for the development of a chinook wind. A chinook is able to remove a snow cover quite rapidly, because it can suddenly increase the air temperature to above freezing. Chinook winds form under two different conditions. One of these is through the flow of air over a mountain or large topographic obstruction. As air is lifted on the windward side of a mountain, it is cooled at the adiabatic lapse rate because of expansion, but considerable heat is also added as liquid water or ice crystals condense or deposit. The net cooling rate on the windward side may be only 7° instead of 10°C per kilometer as occurs as the air descends on the leeward side of the mountain (Figure 3-13). This produces considerable heating, giving rise to a chinook wind that increases the temperature very rapidly in localized areas where the winds reach the surface. Such winds are also very low in humidity on the leeward side of the mountain.

A chinook may also develop after the air has traveled for a considerable distance over higher elevations, across the Rocky Mountains, for example. This air may be heated by radiation with absorption of some energy as it travels across the Rocky Mountains. Then as the

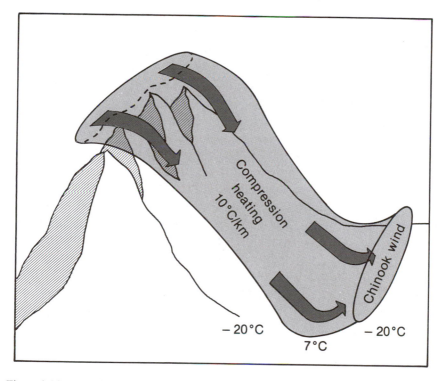

Figure 3-13. The Chinook wind may cause a rapid rise in temperature in localized areas on the leeward side of mountains as compression heating of descending air increases the air temperature and reduces the humidity.

air suddenly descends into lower elevations in the Great Plains, compression heating at 10°C per kilometer results in a chinook wind.

Some examples of the development of chinook winds include very rapid temperature increases at Rapid City and Spearfish, South Dakota. In January, 1911, the temperature rose from −10°C to 6°C, a temperature change of 16°C in only ten minutes. On January 22, 1943, the temperature changed from −20°C to 7°C with a change of 27°C (more than 50°F) in only 2 minutes. Such rapid changes in temperature have been thought in the past to produce psychological effects. In fact, a similar wind that occurs in Germany and in the Alps, called a foehn wind, has been blamed for such great psychological effects that persons committing robberies or other crimes during the foehn wind have not been charged with criminal offenses. Some doctors have suggested that a change in the concentration of positive and negative ions of the air occurs during the chinook wind; this then could be responsible for people reacting in abnormal ways.

PREPARATION FOR WINTER STORMS

Since chinook winds are very localized wind systems, it is much more advisable to plan for dealing with winter storms in some other way. A winter storm car kit is advised for the wise motorist who is anticipating cross-country travel or other travel during the winter. A winter storm car kit should include such items as blankets, sleeping bags, or even boxes of newspapers, since they have considerable insulating effect. Extra clothing such as caps, mittens, and overshoes may come in handy, as may high calorie, nonperishable food. Small cans that have been previously filled with melted wax poured around a wick to form a candle, and a supply of matches could prove to be very valuable for warmth and light. Radios, especially CB radios, may be helpful during times of emergencies. A small sack of sand, a flashlight, a first aid kit, and a shovel are also recommended for a winter storm kit. Additional items which might be quite helpful include booster cables, a tow chain, fire extinguisher, and catalytic heater.

SUMMARY

Blizzards are caused by severe frontal cyclones that produce strong winds, low temperatures, and snow. The longwave cyclone moves slow because the low pressure at the earth's surface is located beneath, and is an extension of, the low pressure center in the upper atmosphere. This type of cyclone may produce severe blizzards because of the extremely low surface pressure, strong winds, and very slow rate of travel. The trough cyclone can also generate blizzard conditions as it forms beneath the trough of the jetstream and is carried along toward the ridge.

Blizzards may produce ice pellets, freezing rain, snow squalls, and blowing and drifting snow, with rain extended southward in the warmer air. Heaviest snow is typically north of the low pressure center about 200 km, while various other types of precipitation occur near the cold front.

Examples of severe blizzards include those of January 1978 when approximately 100 people were killed by a single blizzard. Several severe blizzards also occurred during January of 1977 and 1979 to produce record-breaking temperatures at many locations in the eastern two-thirds of the United States.

The combination of strong winds and low temperatures cause more rapid body heat loss than if winds are light. The wind chill is a measure of the cooling power of the air. Traveling during severe winter storms can be dangerous without a properly equipped vehicle and proper planning. If stranded in a snow storm it is usually preferable to remain in or near the vehicle unless help is within sight.

Chinooks are a very dry wind that cause a rapid rise in air temperature on the leeward side of mountain ranges. The temperature has risen 27°C (more than 50°F) in only 2 minutes.

4
Setting the Stage for
Severe Thunderstorms

The afternoon began with little indication that it was unusual. The hot sun beamed down with such intensity that it seemed to be extracting even more water from the parched earth to make the already humid air even more uncomfortable. As the afternoon progressed a few small cumulus clouds could be noticed in the western sky. One of these seemed to explode as it penetrated rapidly through the atmosphere. As its top began to spread into a mushroom shape, lightning punctuated the darkening sky.

The developing storm began to really show its character as the dark cloud took on a greenish cast and its base grew many small pouchy protuberances. First came the gush of cold air. Shortly afterward sheets of rain joined the blasts of air. As the sound of rain on the roof slackened, as if the storm had passed, a small thump, thump began that was hardly noticeable at first. As the thumps became louder and closer together there was no mistaking the sound of baseball-size hail that intermingled with many smaller hailstones to cover the green grass, and make the world outside unfit for man or beast. Surely the complete silence that soon soaked through the house indicated the end of the storm. But a look at the dark cloud revealed otherwise. A small dark point extending from the southern edge of the cloud showed plainly against the clear western sky. The point grew much larger and extended rapidly downward. As it touched the ground its color darkened noticeably as dark objects spiraled outward away from its base. Its slow forward movement concealed the violence contained within. . . .

ORDINARY AND SEVERE THUNDERSTORMS

The geographical distribution of thunderstorms is shown in Figure 4-1. They form on 100 days per year in some parts of Florida and more than 50 days per year throughout most of the central United States. The largest thunderstorms are most likely to be severe, i.e., contain frequent lightning, and damaging winds or hail. Such large thunderstorms develop in the atmosphere as localized areas of air become unstable from such factors as the absorption of sunlight or contrasting air masses. As a large bubble of air becomes less dense than the surrounding air it is buoyant and rises, perhaps high enough to form a cloud. If other appropriate atmospheric conditions exist, to be described shortly, a severe thunderstorm is produced. The exact location where a severe thunderstorm will develop cannot be forecast since it is dependent upon such factors as the ground beneath

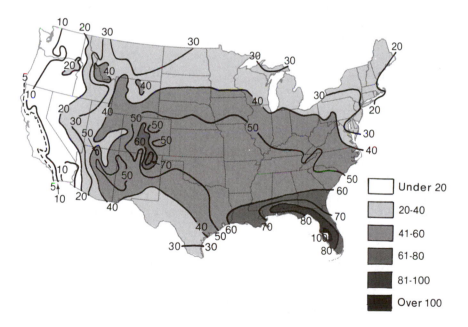

Under 20
20-40
41-60
61-80
81-100
Over 100

Figure 4-1. The average number of days with thunderstorms reaches a maximum in Florida where thunderstorms form on more than one hundred days out of the year. Thunderstorms are common all along the Gulf Coast. A secondary maximum of thunderstorm activity occurs in northeastern New Mexico and southern Colorado where thunderstorms form on more than seventy days out of the year. (After Environmental Data and Information Service, NOAA.)

it, airflow over a hill which initiates an updraft, or converging surface winds that initiate updrafts. Since the location where a specific thunderstorm will develop cannot be forecast we use the next best approach consisting of forecasting the location where the appropriate atmospheric conditions exist for generating severe thunderstorms.

ATMOSPHERIC STABILITY

Severe thunderstorms typically occur on a day with very warm humid air, since the stability of the atmosphere is an important factor. A very unstable atmosphere develops vertical air currents easily; a stable atmosphere, on the other hand, has very few vertical air currents with no mixing in the atmosphere. Forecasters use an index to place a specific number on the stability of the atmosphere for forecasting severe thunderstorms. This index, called the lifted index, is related to the buoyancy of large blobs of air as they rise through the surrounding atmosphere to develop a thunderstorm.

When air changes its height and rises to a higher elevation, it cools at a specific rate called the dry adiabatic lapse rate. The dry adiabatic lapse rate is the rate of cooling of unsaturated air as it expands due to the presence of less atmospheric pressure at higher altitudes above the surface of the earth. A fundamental characteristic of a gas is that expansion produces cooling and compression produces heat. The expansion cooling process occurs at a constant rate as air rises because the atmospheric pressure decreases at a relatively constant rate through the troposphere. This expansion cooling rate is 10°C for every kilometer increase in altitude. Eventually, if the rising air is cooled enough, saturation occurs and water droplets or ice crystals begin to form on particles in the air, producing a cloud.

As a cloud develops, the latent heat of condensation releases 2400 J (575 calories) for every gram of liquid water that condenses (2834 J for 1 gm of ice crystals). The expansion cooling rate is less as the additional heat is added. The expansion cooling rate for saturated air, therefore, consists of expansion cooling (10°C/km) as the air rises into lower atmospheric pressure and latent heat additions as water droplets and ice crystals form in the cloud. The saturated expansion cooling rate is about 7°C/km within the typical cloud layer, but may vary from 5 to 10°C/km. This rate is also called the moist adiabatic lapse rate, although the process is not adiabatic since this term means no heat addition.

The lifted index requires the use of the dry and moist adiabatic lapse rates in order to quantify the stability of the atmosphere for use in severe thunderstorm forecasting. The index is computed by using the average specific humidity (grams of water vapor per kilogram of air) through the layer from the surface to 1 km. Another ingredient is the maximum forecast temperature for the day since this corresponds to the most likely time of severe thunderstorm activity, through the relationship between surface heating and thermal convection. The blob of air having the surface forecast temperature and average specific humidity is assumed to be lifted and cooled at the dry adiabatic lapse rate until it reaches saturation at the cloud base or lifting condensation level. Further lifting proceeds at the saturated expansion cooling rate until the 500 mb level (about 5.5 km) is reached.

The temperature of the theoretically lifted blob of air is compared with the measured temperature at the 500 mb level to obtain the lifted index. The measured temperature is available from weather balloons carrying radiosondes that are sent up twice a day from 93 weather stations in the United States to measure the temperature, humidity and pressure at various heights, including the 500 mb level. Thus, measurements are available for determining the lifted index by subtracting the temperature of the lifted blob of air from the measured temperature of the atmosphere at 500 mb. If the lifted index is negative, it means the temperature inside the blob of air is greater than the surrounding atmosphere; thus the atmosphere is more unstable with severe weather likely. A positive lifted index, on the other hand, means that the atmosphere is stable and the growth of thunderstorms is unlikely.

During the time of the Topeka tornado on June 8, 1966, the lifted index was −6 at Topeka and −4 throughout eastern Kansas and central Oklahoma as two tornadoes formed, one in Topeka and one in central Oklahoma. At 6:00 AM on April 3, 1974, on the day of the worst tornado activity since records have been kept most of the southeastern United States had an unstable atmosphere (Figure 4-2). The unstable region moved northward during the day as 148 tornadoes formed in 11 states east of the Mississippi river.

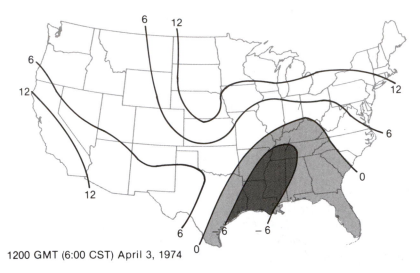

1200 GMT (6:00 CST) April 3, 1974

Figure 4-2. The lifted index is a number computed to represent the stability of the atmosphere. When it is negative, severe thunderstorms are more likely because of the unstable nature of the atmosphere. The lifted index was negative over the southeastern United States on April 3, 1974. More than 100 tornadoes formed out of this unstable air.

MOISTURE TONGUE

Another related factor, important in its own right, is a high dew-point temperature associated with a moisture tongue. A vast source of very moist air exists over the Gulf of Mexico that streams northward into the Central United States when the wind currents are appropriate. This happens when a low pressure area moves into the Central United States, with stronger south winds on the southeastern side of the low as air flows cyclonically around it. This brings the moist Gulf air northward as a tongue-shaped stream of moist air (Figure 4-3). A sharp boundary frequently exists between the Maritime Tropical air from the Gulf and the Continental Tropical air from the Southwest. This boundary is called a dry line and is the western boundary of the moisture tongue. Since the drier air from the Southwest is denser than the moist air (molecular weight of water vapor is only 18, while the average of the gases that form dry air is 28.9) with a similar temperature, the dry line is similar to a cold front with the dry denser air analogous to the cold air. Thus, the dry line is an area where severe weather is frequently observed just as a cold front produces clouds and precipitation.

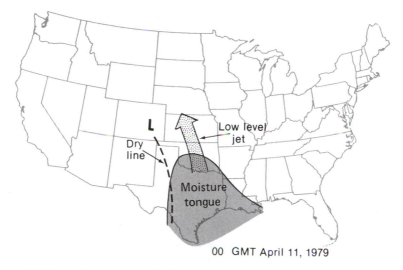

00 GMT April 11, 1979

Figure 4-3. The moisture tongue, dry line, and low level jet are shown here near the time of formation of the very devastating Wichita Falls, Texas, tornado. A tongue of moisture forms as warm moist air is pulled northward around the center of low pressure. The western boundary of the moisture tongue forms a dry line that separates the Maritime Tropical air from the very dry Continental Tropical air. A stream of air called the low level jet also flows northward around the low pressure system. The height of this air stream is about 1 km above the surface.

LOW LEVEL JET

A low level jet commonly flows northward across the Great Plains at a much lower level than the polar jetstream, which flows from west to east at about 12 km (40,000 feet) as previously described. The low level jet is considered a jetstream only in the sense that the airstream has a higher speed than the surrounding air. The low level jet, like the moisture tongue, is related to lower pressure northward with the air flowing around and slightly toward the center of low pressure.

Analyses of the frequency of various wind directions at the height of the low level jet over the Central Plains have revealed the low level jet to be a frequent characteristic of the wind patterns for both summer and winter. During the summer months a band of wind from the Southwest frequently occurs in an area extending from Texas through Oklahoma, Kansas, Nebraska, and Missouri (Figure 4-4). This band of wind frequently exists at intermittent intervals through the Central United States in the winter also, forming the low level jet. The low level jet is important in severe thunderstorm formation, since it provides an air current from an opposing direction and establishes wind shear that contributes to thunderstorm development in a manner to be described in more detail in the next chapter.

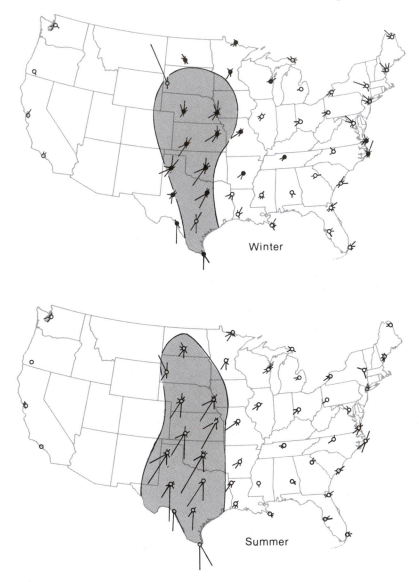

Figure 4-4. Observations of the low level jet have shown that it occurs more frequently in some parts of the United States (shaded area). The direction of the low level jet also varies somewhat between summer and winter and with location. (After *Mon. Wea. Rev.*, 1968)

UPPER AIR INVERSION

Another factor frequently observed when severe weather develops is an upper-air inversion. Inversions are regions in the atmosphere where the temperature increases with height, in contrast to the normal decrease in temperature from the surface upward to about 12 km. The temperature profile prior to severe weather formation typically shows a temperature decrease with height from the surface to about 800 mb followed by a temperature increase with height for a short distance and then a temperature decrease again. The upper air inversion layer consists of the layer where the temperature increases (Figure 4-5).

The presence of an upper air inversion contributes to tornado development by suppressing cloud development until a thermal, or small cloud, is able to break through the inversion layer. This delay frequently allows the resulting thunderstorm to grow much larger than ordinary because the stable inversion layer prevents initial thunderstorm development. The delayed cloud development lets the sunlight continue to warm the ground and lower atmosphere until a

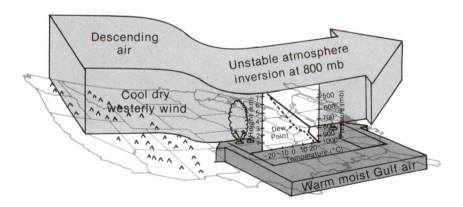

Figure 4-5. As the moist air flows northward across the central Great Plains it encounters the cool, dry, westerly winds descending over the Rocky Mountains. This creates an inversion at the boundary between the two. Small cumulus clouds may be unable to penetrate through this inversion. As a cloud eventually penetrates the inversion, it is able to grow much larger than in a normal atmosphere because of the extreme contrast in moisture and temperature properties between the two layers.

bubble of air is so much warmer than the air above the inversion layer that much more violent thunderstorms form.

MIDLATITUDE CYCLONE

Severe thunderstorm and tornado watches are normally located in the warm air sector of a midlatitude cyclone. The warm air sector is the region of the cyclone south of the warm front and east of the cold front. The surface winds in this region are normally from a south-westerly direction in comparison to those behind the cold front from the north or northwest.

A particular type of midlatitude cyclone frequently produces severe thunderstorms. A midlatitude cyclone that is generated just north of the main axis of the jetstream, yet within the trough of the jetstream is in an ideal position to take advantage of all the various factors associated with the development of strong midlatitude cyclones. These include the cyclonic curvature of the jetstream, cyclonic wind shear just north of the jetstream (cyclonic vorticity) and horizontal divergence. Thus, the midlatitude cyclone that produces severe weather is a hybrid between the trough and longwave cyclones with the strength of a longwave cyclone and the speed of a trough cyclone.

A typical severe thunderstorm producing midlatitude cyclone is shown in Figure 4-6. This cyclone occurred on April 4, 1977, and is situated at the surface just north of the axis of the jetstream and slightly east of the trough. Its position at this location gives it the greatest strength and allows the cold front and dry line to travel at a greater speed as they are pushed forward by the faster winds above. This speed causes more rapid uplift of the warm, moist air, and combines with other strong atmospheric characteristics to increase the likelihood of severe weather.

SQUALL LINES

Although an individual thunderstorm may become severe, the development of a squall line intensifies thunderstorm growth. A squall line is simply a line of thunderstorms, but the thunderstorms interact to reinforce each other. Squall lines develop in the warm air sector of a midlatitude cyclone; frequently along the dry line, since this is an instability zone between the warm, humid air and hot, dry air.

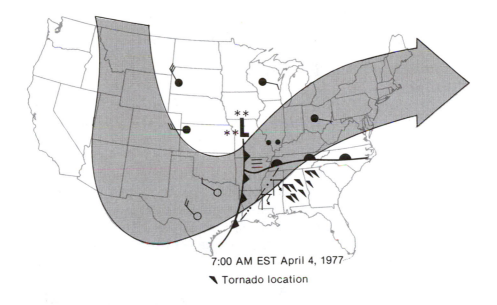

7:00 AM EST April 4, 1977

❧ Tornado location

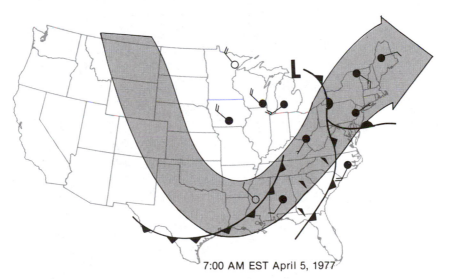

7:00 AM EST April 5, 1977

Figure 4-6. The synoptic map for April 4 and 5, 1977, shows the movement of a frontal cyclone across the central and eastern United States. Squall lines and tornadoes developed within the warm air mass as the frontal cyclone with its associated front moved northeastward. One of the tornadoes that hit Birmingham, Alabama, caused 22 deaths and 15 million dollars worth of damage. Note the path of the frontal cyclone in relationship to the shaded area representing the upper level air flow.

After a squall line has developed, we can pinpoint the most probable location of severe weather much more closely. Prior to squall line development only the large scale atmospheric features are available to forecast the location of possible severe weather, but after the individual thunderstorms form, they can be projected to continue for several hours. The radar screen for Cincinnati is shown in Figure 4-7 for April 3, 1974. Several thunderstorms were developing to the severe stage, as they were carried from the southwest to the northeast by the upper airstream. Each of three thunderstorms developed three different large tornadoes, and a fourth thunderstorm developed a large tornado that struck Xenia, Ohio.

Individual thunderstorms are more likely to be severe if they are located within particular regions of a squall line. One such location is the intersection of a squall line with a cold front. The particular thunderstorm nearest this location frequently reaches the severe stage

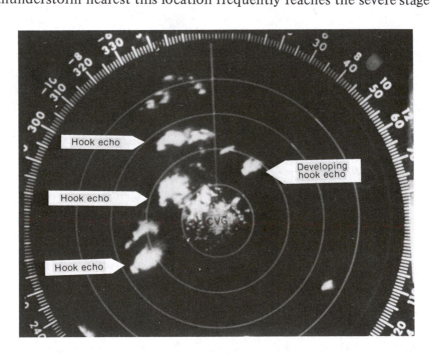

Figure 4-7. The radar screen located near Cincinnati, Ohio, revealed three severe thunderstorms with hook echoes on April 3, 1974. These thunderstorms contained tornadoes that were about 1 km in diameter. A fourth thunderstorm shown on the radar screen toward the northeast of the radar site later developed a large tornado that destroyed much of Xenia, Ohio. The presence of this number of hook echoes at one time is very unusual.

with the help of converging surface winds from the southwest ahead of the front and from the northwest behind the front. These feed air into a developing thunderstorm and intensify updrafts, thus helping the thunderstorm grow faster. Similarly, the intersection of a squall line with the dry line or the warm front is likely to produce a severe thunderstorm. Other locations where severe thunderstorms are likely are near the central bend in a bow-shaped squall line, and at the southernmost end of a squall line, since this thunderstorm is fed by the moist air within the moisture tongue.

WIND PROFILE

The proper vertical wind profile also contributes to the development of severe thunderstorms. The wind direction changes with altitude are very important since such air currents are the life streams of a thunderstorm. The winds, relative to the moving thunderstorm, are of primary importance. The concept of relative winds can be understood by considering a moving automobile. The relative wind for an automobile traveling on the highway at 90 km/h, when the natural wind is calm or against the side of the automobile, is from the front of the car at 90 km/h because of the movement of the car. The car creates a wind equal to its own speed in a direction opposite to its travel. This wind speed can be measured by holding an instrument outside the window. A thunderstorm travels at a typical speed of 50 km/h, producing a relative wind on the front of the storm of this same magnitude. If the air were otherwise perfectly calm, and the thunderstorm was moving at 50 km/h, the winds in the lower atmosphere would strike the front of the thunderstorm at 50 km/h, creating a relative wind.

The strong winds in the upper atmosphere are responsible for the forward motion of the thunderstorm as they push it along from a southwesterly direction, under typical conditions. With a wind speed at the jetstream level of 110 km/h, a large thunderstorm is not blown apart, but travels along as a unit because of air currents and rotation within the thunderstorm. With a light surface wind from the southeast and winds against the upper part of the thunderstorm at 110 km/h, to produce thunderstorm movement of 50 km/h, the relative wind speed at the back of the thunderstorm pushing it along would be the difference between the movement of the thunderstorm and the speed of the upper level winds from the southwest, or 60 km/h. The winds at the front of the lower part of the storm would be from the northeast and almost as strong, 50 km/h. The relative winds are, therefore, much different from the actual measured winds above a particular location.

The proper wind profile for the development of severe thunderstorms consists of strong opposing relative winds with considerable wind shear (Figure 4-8). Wind shear exists in the atmosphere if the winds in different layers are from different directions or if a moving thunderstorm creates winds from opposing directions. Strong winds above lighter winds or winds from opposing directions cause horizontal roll clouds that may be tilted into vertical positions to become the rotating cores of severe thunderstorms. Strong wind shear is also required to support the double vortex internal thunderstorm structure, after it has formed. Thus, a strong wind shear environment is conducive to severe thunderstorm formation.

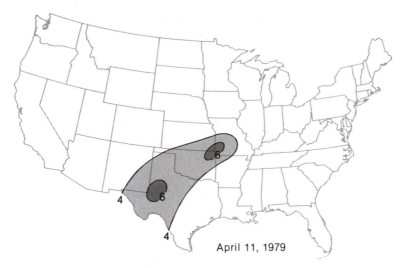

April 11, 1979

Figure 4-8. The wind shear relative to a moving thunderstorm may provide information on where tornadoes are likely to develop. A wind shear index can be computed by counting the number of layers above 850 mb that contain winds approximately opposite in direction and equal in magnitude to the low level winds below 850 mb. The wind shear index showed that 4 of the 6 layers oppose the low level winds on April 11 at 00 GMT, at about the time of the Wichita Falls tornado. Other tornadoes developed in Texas and Oklahoma on this date. The wind shear is another of several factors that are important in tornado development.

UPPER AIR DIVERGENCE

Atmospheric regions that contain upper air divergence are very favorable for severe thunderstorm formation. Upper air divergence occurs as airstreams spread apart and represent a region with net outflow of air (Figure 4-9). Such regions in the upper atmosphere tend to suck the air up from beneath, thus initiating or intensifying updrafts in thunderstorms located beneath such areas.

Two of the worst tornado outbreaks in history occurred on April 3, 1974, and April 11, 1965. On April 3, 148 separate tornadoes caused more than three-fourths of a billion dollars worth of damage with thousands of people injured and 265 people killed over an 11 state area. Upper air divergence was a major factor in this widespread severe thunderstorm activity. Many of the tornadoes were quite large (1 km in diameter at the base). This was very unusual but possible since all the atmospheric conditions were present for the formation of large thunderstorms. On April 11, the upper air divergence was not quite as pronounced and the trough of the jetstream was less curved than in

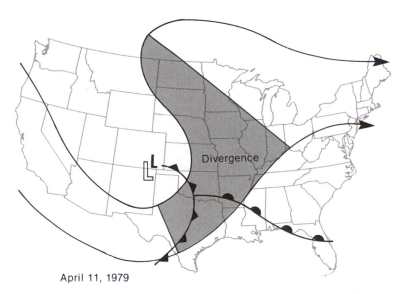

April 11, 1979

Figure 4-9. Diverging air currents above the Earth's surface are important in generating severe thunderstorms beneath since updrafts located beneath areas of diverging air are likely to be much stronger. Considerable divergence existed over the central United States at the time of the Wichita Falls tornado. The tornado activity was confined to Texas and Oklahoma, however, because of other tornado formation factors.

the previous case, but considerable divergence was also present that contributed to the formation of many large thunderstorms. Thus upper air divergence contributes to severe thunderstorm development by producing a vacuum effect in the upper atmosphere that contributes to strong updrafts.

JETSTREAM

The jetstream is a major factor in the development of severe thunderstorms. The jetstream is important because of: 1) the Conservation of Angular Momentum, discussed previously, with available energy transferred from the large circulation down to smaller circulation, thus providing a source of energy for the severe thunderstorm and tornado; 2) the influence it exerts on the midlatitude cyclone. As the jetstream goes through cycles (first as a west to east wind, then to a stream of air that meanders north and south, and eventually forms loops that are cut off before the airstream returns to a west-east flow again) it helps to generate areas of low pressure (midlatitude cyclones) near the surface (Figure 4-10). These then develop a cold

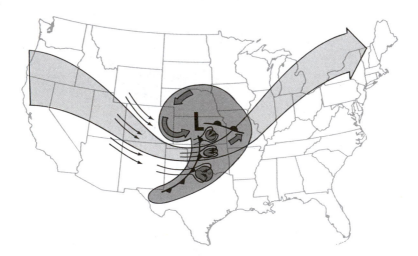

Figure 4-10. The jetstream not only helps to form the frontal cyclone within its cyclonic bend but it also feeds energy into individual thunderstorms by the strong air flow around them. Its cyclonic curvature also strengthens cyclonic vorticity (circulation) within an individual thunderstorm.

front, and perhaps a dry line along with the various other atmospheric conditions that contribute to severe thunderstorm activity. On a smaller scale, the strong winds supplied by the jetstream around the thunderstorm are important in initiating and maintaining rotation within the severe thunderstorm; thereby, feeding energy directly into an individual thunderstorm.

In addition to the role of the jetstream in generating midlatitude cyclones, it also determines the speed and direction of movement of the midlatitude cyclone. The typical path of a midlatitude cyclone is from the southwest to the northeast because of jetstream flow in this direction. If on a particular day a very pronounced curvature exists in the jetstream producing a northward bound jetstream, this causes a low pressure center located beneath it to move in this same direction. If the speed of the jetstream is greater, it will move the cyclone faster and more energy will be available for smaller circulation.

SEVERE THUNDERSTORM WATCHES

A severe thunderstorm watch is issued by the National Severe Storms Forecast Center located in Kansas City, Missouri, on the basis of a combination of all the appropriate atmospheric conditions just described. The intersection of the moisture tongue and low level jet with a line projected downward from the axis of the jetstream is a common location of severe weather. This is typically located within the warm air sector of a midlatitude cyclone, and also corresponds to a region of upper level divergence and cyclonic curvature of the jetstream (Figure 4-11). The lifted index is frequently more negative farther southward in the warmer air. However, severe thunderstorm activity is most likely where the most divergence exists above the moisture tongue and low level jet if an upper air inversion is present. With these conditions, squall lines develop in advance of the cold front along the dry line. The area described in a severe thunderstorm forecast is frequently a rectangle (Figure 4-12). It is located in the warm air sector of a midlatitude cyclone and may be bounded to the north by the warm front and to the west by the cold front. Since the whole weather system is moving, the line of most likely activity continually moves toward the east; therefore, a rectangle is needed to describe the forecast area of severe weather for more than an instant in time.

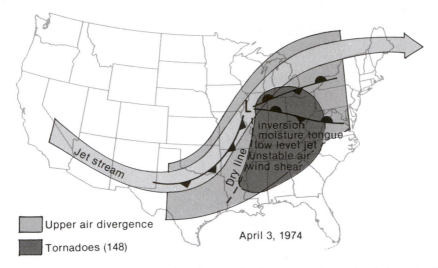

Figure 4-11. The geographical location of severe thunderstorms and tornadoes is determined by a combination of the various atmospheric conditions that help generate them. If all the individual atmospheric elements are present over a large area, as on April 3, 1974, many tornadoes can develop.

You can frequently tell when the danger of severe thunderstorms is past before the watch is officially ended by noting the presence of the cold front. If the wind changes direction from the southwest to the northwest, and the temperature decreases, this indicates the front has passed, with a decreased likelihood of severe weather. Additional "inside information" is available after squall lines have developed within a severe thunderstorm watch area, since the whole watch area does not normally have thunderstorms. Existing thunderstorms can be tracked by radar, or visually, and are likely to move toward the northeast because of the upper level winds. Therefore, the area most likely to have severe weather is pinpointed much more closely than the large rectangle describing the entire severe thunderstorm watch area. In some cases, a severe thunderstorm watch is issued and no clouds develop. Additional triggering mechanisms such as topographic features, and uneven heating at the earth's surface are sometimes needed to start thunderstorm development.

The impact of the ordinary severe thunderstorm watch is weakened because of the size of the area. Ordinarily, 90% of the watch area may not even have cloud cover, and only one or two thunderstorms within the area develop hail, high winds, or a tornado. This makes

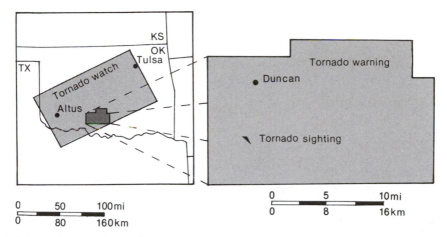

Figure 4-12. Tornado watch areas are issued on the basis of the presence of appropriate atmospheric conditions for tornado formation; thus, they cover a large geographical area. Tornado warnings cover only a few counties and are generally issued after a tornado has been sighted.

the forecast seem much worse than it really is to the many people who do not see a cloud. In addition, much room for improvement of our forecasting techniques exists, since less than one-third of the tornadoes that develop are in watch areas. More of the large ones are within forecast areas, however, since over 50% of those that cause deaths are within forecast watch areas.

The atmospheric conditions required for severe thunderstorms vary somewhat with geographical location. In the southern United States the atmosphere becomes so humid and unstable that severe thunderstorms can grow in the absence of some of the other specific atmospheric conditions, such as the jetstream, that are otherwise required for their formation. Thus, the relative magnitude of the various atmospheric conditions is important, which makes the problem much more complex than it would be if we were dealing only with the presence or absence of the various atmospheric conditions.

SUMMARY

Severe thunderstorms form when several different atmospheric conditions coincide. The typical severe thunderstorm day begins with sunny skies that allow sunlight to warm the earth and lower atmosphere. A warm, moist air current flows northward from the Gulf of Mexico toward and around a low pressure center, resulting in a low level jet and a moisture tongue. Strong westerly flow in the upper troposphere, provided by the jet stream, brings cooler air over the Rocky Mountains. This increases the instability of the atmosphere, producing a negative lifted index and creating an inversion layer.

As the midlatitude cyclone strengthens (because of its position just north of the jet stream axis in a region of cyclonic vorticity, cyclonic wind shear, and horizontal divergence) its forward movement and circulation help generate thunderstorms in advance of the cold front along the dry line. As particular thunderstorms within a squall line encounter favorable environments for growth, such as intersection with the dry line or a weather front, they grow more rapidly and become severe.

Forecasting of severe thunderstorms is the responsibility of the National Severe Storms Forecast Center in Kansas City. Severe thunderstorm watches are issued for any geographical region in the United States where atmospheric conditions are favorable for the growth of severe thunderstorms.

5
Nature of Severe Thunderstorms

To an avid hot-air balloonist the prospect of taking an instrument package up to the base of the large normal looking thunderstorm seemed only mildly adventuresome. The balloon rose slowly near the front of the cloud. As the balloon approached the thunderstorm a stream of air accelerated us into a notch-shaped opening. The ascension rate increased as the cloud droplets within the updraft engulfed us and began to make us feel wet and cold. As the air temperature dropped below freezing, we soon became concerned about the coating of ice that began to form everywhere. From gravity we should have been falling due to the ice formation, but our sensations told us we were still rising rapidly. Air currents seemed to be whirling in all directions around us.

The pulsating light from lightning discharges, dangerously near, revealed hailstones carried by the whirling air currents collecting near our feet. As the temperature dropped to 40 below zero, our concerns compounded. The extreme wind gusts were ready to tear our craft apart at any moment, and the wind and cold were beginning to chill us through our heavy clothing. At that moment the bottom seemed to fall out of the sky as we were caught by a downdraft. We knew we were much lower when liquid water droplets pelted our faces. As we emerged from the cloud we realized that we were traveling downward at such a speed that we would be extremely lucky if we weren't dashed to bits against the ground. . . .

THUNDERSTORM DEVELOPMENT

Thunderstorms have their origin in small cumulus clouds that begin to grow vertically through the troposphere. While cumulus clouds are quite common, only a very small percentage become large cumulonimbi and reach the severe thunderstorm stage. This stage is reached when the thunderstorm develops frequent lightning, accompanied by locally damaging winds or hail of 2 cm diameter or larger. Damaging winds are defined as sustained or gusty surface winds of at least 97 km/h. Severe thunderstorms also produce tornadoes, but are classified as severe thunderstorms when no hail or tornado is present if they generate locally damaging winds and frequent lightning.

The ordinary thunderstorm has a lifetime of less than one hour, because it grows through an atmosphere where the winds increase with height and tend to shear off its top as it builds. The thunderstorm that reaches the severe stage, however, must develop an internal structure that allows it to thrive on the strong upper level winds, since its lifetime is several hours.

When the appropriate atmospheric conditions are combined, as in a severe thunderstorm watch rectangle as described in the previous chapter, a thunderstorm may grow to the severe stage from an isolated thunderstorm, squall line thunderstorm, or multicell thunderstorm. The single isolated "supercell" thunderstorm develops as a cloud breaks through the inversion layer and grows rapidly through the stronger winds above (Figure 5-1). This frequently requires more than one trial as one cloud is eroded away by the strong winds, and a second rising turret encounters the more humid air left from the decayed cloud and is able to grow more rapidly through the less harsh environment.

As thunderstorms develop in a row, forming a squall line, each thunderstorm influences the adjoining one through the effect of the gust front and blockage of some of the upper level winds. The gust front is a miniature cold front created at the ground beneath the leading edge of a thunderstorm as a cold downdraft spreads outward on striking the ground (Figure 5-1). This may reinforce the updraft of an adjoining thunderstorm by pushing warm, moist air upward. The squall line gives some degree of blockage of upper level winds. This makes it easier for new cells to grow downwind from the squall line.

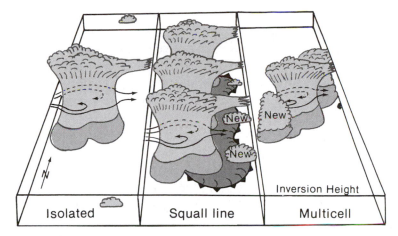

Figure 5-1. Thunderstorms may reach the severe stage as isolated thunderstorms, within squall lines, or as multicell thunderstorms.

Multicell thunderstorms may also reach the severe stage. This type of thunderstorm grows in a cluster of perhaps three cells, where the downwind cell is decaying, the central cell is mature, and the upwind cell is new and developing (Figure 5-1). The central mature thunderstorm may become a severe thunderstorm.

Regardless of the growth pattern, supercell, multicell, or squall line, there is evidence that the internal structure of the severe thunderstorm is similar in each case. The severe thunderstorm develops an organized internal structure to allow it to survive longer and grow larger than the ordinary thunderstorm.

AIR CURRENTS WITHIN THUNDERSTORMS

Simple thunderstorm models developed in the 1800s by various scientists suggested the presence of updrafts and downdrafts in severe thunderstorms as a step in explaining such observed features as lightning, hail, and tornadoes. It has only been in the 1970s, however, that remote sensing techniques, specifically dual Doppler radar, have been sufficiently developed to give information on the wind currents within a severe thunderstorm. These have shown such features as counter-rotating vortices inside the severe thunderstorm. The existence of such a double vortex structure was suggested from the theory of air flow around barriers a few years before such measurements were available for verification.

Simple visual observations of a severe thunderstorm in action provide some clues to the internal structure. Large thunderstorms extend from their cloud base at perhaps 1 km to heights as great as 18 km. The wind environment surrounding them typically includes light winds near their base and strong winds of perhaps 150 km/h at the jetstream near 12 km. In spite of this strong shear, the large thunderstorm is able to withstand these high winds to avoid being tilted over or sheared off as happens when small thunderstorms start to grow. In addition to visual observations, satellite and other photographs of large thunderstorms show that they maintain a fairly vertical position (Figure 5-2).

The most severe thunderstorms stay in a vertical position for several hours. This tells us that the thunderstorm must have the appropriate internal wind currents to counterbalance the strong external winds which would otherwise penetrate the thunderstorm and tear it

Figure 5-2. Severe thunderstorms are able to grow very quickly through a strong wind shear environment. Their tops spread out to form anvils as they penetrate into the stratosphere.

apart. Thus, a large persistent thunderstorm must represent a barrier to the environmental wind flow. The equations describing wind flow around a barrier can be used to study the required air currents inside the thunderstorm. Their application reveals that the thunderstorm would form a barrier to the environmental winds if it contained a double vortex circulation as shown in Figure 5-3. A cyclonic rotating cell in the southern half of the thunderstorm and anticyclonic rotation in the northern half produce a strong east to west flow through the center of the thunderstorm to counterbalance the environmental winds striking the thunderstorm from the west. In addition, the two counterrotating cells cause the environmental air to flow around the sides of the storm with very little interference, since the winds on the sides of the storm are moving in the same direction as the surrounding flow.

The presence of cyclonic rotation within a large thunderstorm is being used to develop a method of forecasting tornadoes since Doppler radar will detect the rotation as it develops deep within the thunderstorm. Since this forms as much as 30 minutes before tornado touchdown it represents an important new tornado warning method.

Since many thunderstorms complete their life cycle without developing into a severe thunderstorm with an organized internal struc-

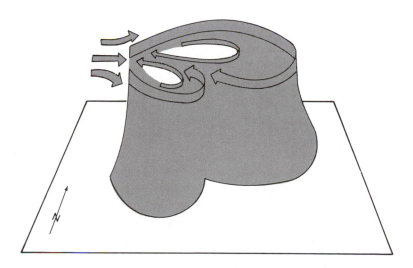

Figure 5-3. A double vortex internal flow pattern is most efficient in allowing a thunderstorm to withstand the strong winds surrounding the thunderstorm. Without such a flow structure the environmental winds would penetrate through the thunderstorm.

ture, the mechanism for the development of a double vortex structure that allows a storm to persist for several hours is of special interest. A double vortex internal structure may originate simply from air
flowing around a thunderstorm, interacting with the sides of the storm,
and creating a cyclonic vortex in the southern part of the storm with
an anticyclonic vortex in the northern part. It is also likely, however,
that the vertical wind shear originates the double vortex structure.
As a cumulus cloud grows higher it encounters stronger winds that
can be observed to blow the top off small clouds. As the cloud grows
higher, the extreme wind shear develops horizontal vortices, similar
to roll clouds. Just below the jet stream the wind shear is sufficient
to form a horizontal vortex strong enough and long enough for the
ends of the vortex to tilt downward into a more vertical position (Figure 5-4). The southern end of the vortex rotates cyclonically and the
northern end rotates anticyclonically as they become integrated into
the thunderstorm structure. Such rotation is further enhanced by
the airflow around the sides of the thunderstorm, thus forming a
stable internal flow pattern that is compatible with the strong surrounding airflow.

In the mature stage, the internal flow of a double vortex thunderstorm allows it to exist in a vertical position, since the winds inside
the thunderstorm now oppose the strong environmental winds com-

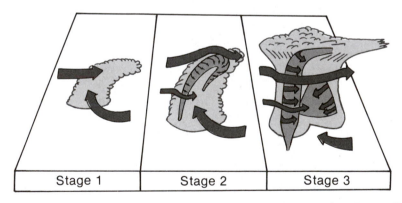

| Stage 1 | Stage 2 | Stage 3 |

Figure 5-4. The internal structure of severe thunderstorms develops as the rising, small updraft forming the thunderstorm begins to block the surrounding air flow. The double vortex
structure develops as horizontal circulation originated because of the strong vertical wind
shear is tilted into the vertical. The cyclonic circulation within the southern half of a thunderstorm and anticyclonic circulation within the northern half may also increase because of
surrounding air flow around the southern and northern sides of the thunderstorm.

ing from the opposite direction at the upper backside of the thunderstorm. The counterrotating vortices also help draw additional warm, moist air into the front of the storm at low levels as shown in Figure 5-5. Thus, the storm has achieved internal flow patterns that allow it to survive, and, in doing so, has also created the mechanisms for the production of damaging surface winds, lightning, hail, and tornadoes.

Thunderstorm models such as the double vortex model, allow us to consider the thunderstorm as a whole; something that is not always possible from other approaches. A closer look at the detailed flow patterns inside a severe thunderstorm helps explain the weather associated with it. As the strong environmental winds flow against the back and around the sides of the thunderstorm, they propel the thunderstorm along and create a wind at low levels that strikes the thunderstorm from the front (east side). This air goes into the thunderstorm and rises, since it is warm, humid air that is light and buoyant. The air coming into the front of the thunderstorm rises, because of its temperature and water vapor content, until it strikes the back side of the thunderstorm, where it is forced to spiral around cyclonically on the southern side and anticyclonically on the northern side of the cell.

Raindrops begin condensing within the interior of the storm as the warm air rises within the central part. Most of the rain falls from the northeast section of the thunderstorm, since the upper level winds blow the raindrops downwind into the anticyclonic rotation. The anticyclonic rotation within the thunderstorm does not develop as much strength as the cyclonic vortex because of the obstruction of large masses of rain. The anticyclonic vortex is also weaker than the cyclonic vortex because of the large scale cyclonic curvature of the jetstream. As cyclonic vorticity and curvature are fed into the cyclonic vortex (also called mesocyclone), its rotation becomes much more pronounced than the anticyclonic rotation. This leaves the southern part of the thunderstorm with a very strong counterclockwise or cyclonic rotation that is further enhanced as it develops strength enough to centrifuge raindrops out of the vortex to eliminate their obstruction to airflow. As the cyclonic vortex grows it begins to simulate a vertical tube extending from the lower part of the thunderstorm to its top. The interior of the vortex tube then develops an updraft because of the higher surface pressure and the influence of the jetstream flowing over the top of it. This dynamic updraft through the center of the cyclonic rotation, combined with the horizontal rotation, eventually extends down to the ground as a tornado. This has led to the development of "tornado signatures" within thunderstorms that provide as much as 30 minutes warning prior to the development of a tornado on the ground.

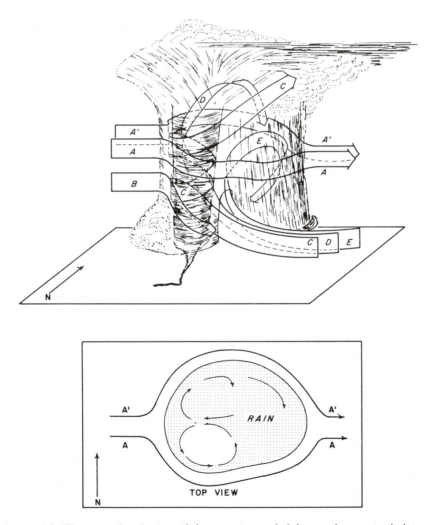

Figure 5-5. The severe thunderstorm thrives on a strong wind shear environment as its internal flow pattern becomes organized into a pattern that matches the surrounding air flow. The warm moist air flowing into the front of the thunderstorm at low levels counterbalances the air flow in upper levels and forms the internal double vortex structure. (After Eagleman and Lin, *Journal of Applied Meteorology*, October, 1977.)

The outer shell of the cyclonic vortex is generated by the thermal updraft, the warm, moist air traveling from low levels into the front part of the storm that rises and collides with the environmental winds at the back of the thunderstorm, thus feeding the cyclonic vortex. Within the cyclonic vortex, the smaller dynamic updraft becomes quite strong due to the pressure difference between its top and bottom, and the jetstream flow over it. The anticyclonic vortex does not become strong enough to simulate a tube, thus it does not develop a central updraft. On the contrary, the effect of falling rain is great enough to initiate a downdraft through the core of the anticyclonic rotation. The major rain area, therefore, occurs in the leading northeast edge of a thunderstorm, while if there is a tornado, it extends from the southern edge of the cloud.

It has been observed that once a thunderstorm has developed the appropriate internal structure to generate a tornado, it approaches more steady-state conditions and has a longer lifespan. Also, because of the more steady-state internal structure, a thunderstorm that develops one tornado is likely to develop another one after the first one dissipates.

The downdraft within the rain area of the anticyclonic vortex is typically strong enough to reach the ground. As it strikes the ground it spreads outward ahead of the approaching rain in the leading part of the thunderstorm to form the gust front. This gust of cooler air helps feed the thermal updraft into the storm by starting the warm surface air upward (Figure 5-6). This makes the thunderstorm even more efficient after it reaches the supercell stage. Thus, the air currents reinforce each other, with additional reinforcement as raindrops fall through the cloud and evaporate some water to cool the air and contribute further to the downdraft, since cold air is denser and accelerates downward.

Measurements of the internal structure of thunderstorms have only recently been possible from dual Doppler radar. The Doppler effect is used to measure the speed of raindrops in the thunderstorm as they reflect radiowaves back to a receiver. The use of two or more Doppler radar sets aimed at the same thunderstorm allows the three dimensional airflow patterns within a thunderstorm to be reconstructed.

Dual Doppler radar data were collected from a thunderstorm on June 8, 1974 near Oklahoma City for several different times throughout the life cycle of this tornado-producing thunderstorm. The Dop-

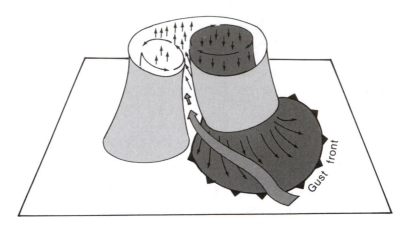

Figure 5-6. The cold air outflow through the rain area in the northeast part of a thunderstorm spreads out ahead of the storm as it strikes the ground. This generates a gust front that is somewhat similar to a miniature cold front. This helps to push the warm moist inflow air current into the thunderstorm.

pler data were collected as thousands of individual data points for each of the two radar sets located in different places. After considerable analysis these were displayed as horizontal slices through the thunderstorm as in Figure 5-7. A tornado developed at 14:23, just 12 minutes after the first data set. The Doppler data for this time shows the existence of the hook echo, and a fairly well developed double vortex structure extending vertically for some distance through the thunderstorm. The double vortex structure was much less developed at 14:11 prior to tornado formation although the cyclonic vortex is evident. The double vortex structure developed at 5 km and extended up to 7 km. The hook echo was generated by the cyclonic flow of air from the southern part of the thunderstorm that carried some of the rain around with it to give the hook-shaped radar echo.

THUNDERSTORM MOVEMENT

The movement of thunderstorms is determined primarily by the upper level winds with some influence from thunderstorm rotation. If the thunderstorm is not rotating, and the average winds from the surface to the top of the thunderstorm are from the southwest, then the thunderstorm will move with the prevailing winds. On the syn-

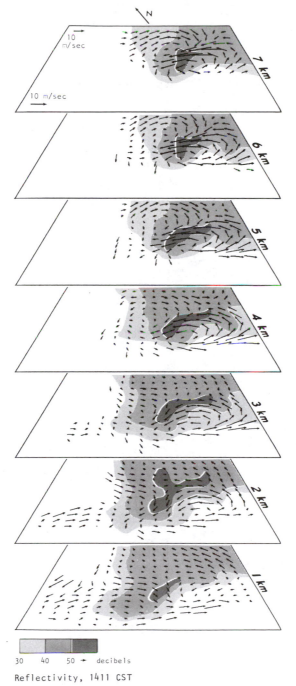

Reflectivity, 1411 CST

Figure 5-7. Dual Doppler Radar data can be used to measure the air currents within a thunderstorm. Such data from a tornado producing thunderstorm show the cyclonic circulation in the southern part of the thunderstorm at various levels. The anticyclonic circulation is also evident on the slices from 4 km to 7 km. Precipitation is concentrated between the double vortex structure at these heights as indicated by the darker shading corresponding to greater radar reflectivity. A tornado extended from this thunderstorm at about the time of these measurements. (After Eagleman and Lin, *Journal of Applied Meteorology*, October, 1977.)

optic map for June 8, 1974 (Figure 5-8) the weather patterns corresponding to the dual Doppler radar data, shown previously, show a
typical upper airflow from the southwest.

The rotation of a thunderstorm may affect its direction of travel if
the rotation is very pronounced. Cyclonic rotation produces thunderstorm movement to the right, when looking downstream with the
mean wind, whereas anticyclonic rotation produces movement to the
left. If the mean wind is from the west, and a large thunderstorm has
considerable cyclonic rotation, then the moving force is the average
wind from the surface to the top of the thunderstorm, which is carried
along with the winds, except for some deviation to the right of the
mean wind (Figure 5-9). This rotation force is called the magnus
force, and it deflects a thunderstorm just as a baseball pitcher spins

Figure 5-8. Thunderstorm movement is frequently from the southwest because of the upper-
air currents, as shown here for June 8, 1974, at the time of tornado formation in Kansas and
Oklahoma. The cloud in central Oklahoma was analyzed in detail by dual Doppler radar as
shown in the previous figure. The large arrow shows the typical jetstream flow pattern
which is south of the center of low pressure to carry thunderstorms located beneath it toward
the northeast. (After Eagleman and Lin, *Journal of Applied Meteorology,* October, 1977.)

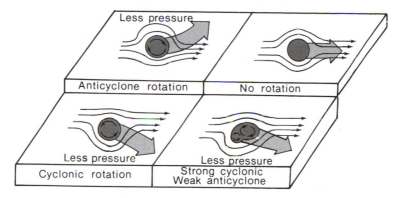

Figure 5-9. Thunderstorm movement is influenced by the upper level winds that carry it along and by its rotation. Anticyclonic rotation is accompanied by deviations to the left of the average winds while cyclonic rotation will cause a thunderstorm to travel to the right of the average air stream. A double vortex thunderstorm with stronger cyclonic rotation will deviate slightly to the right of the main air current that is pushing it along. The reason for these deviations is a reduction in pressure on one side of the thunderstorm because of the rotation.

the ball to throw a curve ball. The flow of air around the spinning thunderstorm produces greater pressure on one side, and this causes it to curve.

Since most severe thunderstorms have some cyclonic and some anti-cyclonic rotation, the rotations tend to cancel except that the cyclonic rotation is usually stronger, resulting in thunderstorm movement from the southwest in the direction of the mean wind, with some deviation to the right of the mean wind. The synoptic situation on June 8, 1966, during the Topeka tornado shows winds from the southwest (Figure 5-10). For the major tornado outbreak occurring on April 3, 1974, the severe weather action was to the east of the trough in the upper level winds as they moved from the southwest; wind speeds were high in this case, so that the average speed of the thunderstorms was about 90 km/h. On April 11, 1965, about 50 tornadoes were generated in a similar situation with movement from the southwest to the northeast. The Topeka thunderstorm moved slightly to the right of the mean winds similar to many of the thunderstorms on April 11, 1965, which traveled along at about 20 degrees to the right of the mean winds. Typical deviations of severe thunderstorms are from 10 to 30 degrees to the right of the mean winds. This indicates

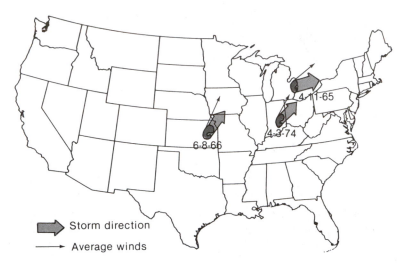

Figure 5-10. The direction of thunderstorm movement is shown here for three major tornado days. Thunderstorm movement was to the right of the average wind in each case. A few thunderstorms have also been observed to move to the left of the average winds.

the existence of cyclonic rotation to deflect them slightly from the mean wind direction.

Thunderstorms frequently slow down as they develop into the severe stage. After the Topeka tornado, the next tornado that caused damage in excess of $100,000,000 occurred on May 11, 1970, in Lubbock, Texas, from a very slow moving thunderstorm. The synoptic situation in this case did not immediately appear to be typical of the Central Great Plains tornado, since the jetstream was some distance northward and no cold front existed near Lubbock. The air was quite humid, however, and the low level winds were opposite in direction and similar in speed to the weak upper level westerly winds. Thus, a stationary thunderstorm could acquire the internal structure to form a tornado due to the opposing winds. The severe thunderstorm that developed over Lubbock showed very little movement for more than two hours because of these generating factors. Such thunderstorms require even more careful analysis for forecasting than more typical severe thunderstorms.

The thunderstorm over Lubbock developed a hook echo on the radar screen indicating that it had formed the double vortex internal structure. Upper level winds of only 30 km/h may be sufficient to drive the internal winds after the thunderstorm has penetrated through them. If the thunderstorm has little or no movement, then the actual winds become the relative winds and low level wind directions must be appropriate to feed the storm with inflow from the southeast feeding the thermal updraft and opposing the winds in the upper part of the storm.

The thunderstorm speed of travel is, thus, related to the average winds from the bottom to the top of the thunderstorm, with an additional influence from the amount of rotation of the thunderstorm. Thunderstorms with considerable internal rotation are pushed along at a slower rate than those with less rotation, since a well developed double vortex structure forces the approaching winds around the outside of the storm with minimum friction and a corresponding loss of power to push the storm along. If the thunderstorm rotation is at a greater speed than the relative wind around it, it can move upstream for a short time period as the greater rotational winds propel it against the surrounding winds. This can continue for only a short time, since the thunderstorm must derive its energy for rotation from the surrounding air stream, but is possible particularly when the winds are changing and the rotation within the thunderstorm has reached its peak. In general, however, the thunderstorm moves along at a rate determined by the environmental winds with a typical speed of 50 km/h. It has been observed that the most severe thunderstorms slow down at about the time they form a tornado. This behavior is intriguing, as if they were stopping to liberate the tornado. Actually, this is what is happening since the increasing circulation, which generates the tornado, also reduces the ability of the environmental winds to push the thunderstorm along.

Another observation of severe thunderstorms is that occasionally a single thunderstorm will split into two different cells. When this occurs the southern most cell turns toward the right, while the one on the north turns toward the left (Figure 5-11). This is another indication of cyclonic rotation in the southern part of the cell and anticyclonic rotation in the northern part. This may happen repeatedly with one cell splitting into two cells, then several hours later, these cells split again. In this way a single thunderstorm can generate several additional thunderstorms that are usually severe and can produce tornadoes.

Some indications point to the existence of preferred paths for thunderstorms. One of these seems to be toward large cities such as

St. Louis, Kansas City, and Oklahoma City. The localized area where
thunderstorms originate may be unique with a topographic feature
such as a hill or darker field that initiates upward movement of air
sufficiently to start the development of a thunderstorm. This then
triggers thunderstorm development in the same spot time after time.

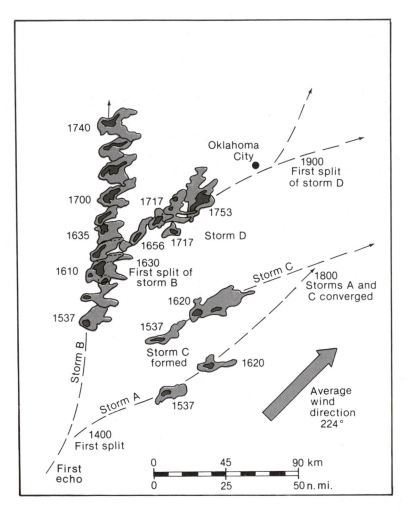

Figure 5-11. Thunderstorms have been observed to split, as shown by these radar cells. When
this happens, the cell on the right tends to deviate to the right of the average wind direction
while the one on the left deviates to the left of the average wind. Such paths would be expected if the original thunderstorm that split had a double vortex internal structure. (After
Charba and Sasaki, NSSL Tech. Memo 41, 1968.)

There are indications of this if you look at tornado paths across Arkansas, for example, where there are definite ridges with higher topography. The paths of tornadoes are parallel to these features although this does not prove a cause and effect relationship.

There may be some effect from cities themselves on thunderstorm formation and tornadoes. We have measured the urban heat island in and around cities of various sizes and found that for Lawrence, Kansas (population 50,000), there is on the average a 2°C heat island over the city. A heat island of 3°C exists in Topeka (population 125,000), and a heat island of 4°C is common in Kansas City (population more than one million). If the air over the city is hotter, this tends to initiate updrafts over the city. Thunderstorms approaching a particular city with rising air above it, would have some tendency to move toward the rising air and warmer temperature.

The lifetime of a thunderstorm is usually a couple of hours. Figure 5-12 shows a thunderstorm that originated near Cincinnati, Ohio,

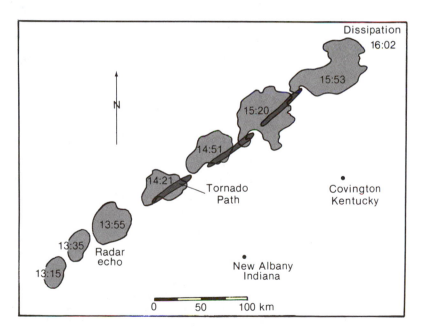

Figure 5-12. The radar echoes shown here reveal the life history of a particular severe thunderstorm that occurred on April 3, 1974, west of Cincinnati, Ohio. After more than an hour's growth, the thunderstorm reached the severe stage and developed a tornado at about 14:21. Within the next one and one half hours it formed two additional tornadoes before dissipating.

on April 3, 1974. It started as a very small cell at 13:18; by 14:21 it was developing a hook shape; by 15:53 it still had a hook shape and produced tornadoes for about an hour and a half; by 16:02 the echo was much smaller and weaker. Thus, this severe thunderstorm lasted for about three hours. This particular thunderstorm produced three different tornadoes, all quite large and severe, during the time from 14:21 to 15:53.

SUMMARY

Severe thunderstorms produce frequent lightning, and locally damaging winds or hail. They may also contain one or more tornadoes. A severe thunderstorm develops as a single supercell thunderstorm, squall line thunderstorm, or multicell thunderstorm. In each case, the severe thunderstorm generates an organized internal structure that allows it to last longer and grow larger than the common variety of thunderstorms.

Large thunderstorms are observed to grow vertically through strong wind shear environments, and form a barrier to the surrounding airflow. Air currents form a double vortex inside the severe thunderstorm with a cyclonic vortex in the southern part, and an anticyclonic vortex in the northern part of the thunderstorm. As the cyclonic vortex dominates, due to general cyclonic curvature of the atmosphere and obstruction from rain in the anticyclonic vortex, it begins to simulate a tube extending through most of the thunderstorm. Its extension to the surface produces a tornado.

Techniques are being developed for using Doppler radar to see the cyclonic rotation within the severe thunderstorm as much as 30 minutes before it is strong enough to extend to the surface as a tornado.

The movement of severe thunderstorms is in response to the mean winds through the atmosphere with some influence from the rotation of the thunderstorm. A strong double vortex structure allows the thunderstorm to slow down because of less friction as the airstream flows around it. If the cyclonic rotation is stronger than the anticyclonic, the thunderstorm will move to the right of the mean winds.

Thunderstorms that split into two cells travel in separate directions, with the southern cell moving to the right and the northern cell to the left of the mean wind. This movement is to be expected from the double vortex structure of a severe thunderstorm.

Severe thunderstorms typically last for an hour or two with the longest lasting storms only a few hours. They may follow the same path more than once because of unique topographic features.

6
The Strongest Storm on Earth

As we drove through the blinding rain we both sensed that we should have gone to our basement instead of chasing the storm. As we broke out of the rain area traveling southward, we saw the tornado ahead of us and to our right. We thought that perhaps we could cross in front of it and circle behind out of danger. Our plan seemed to be working, and gave us a view that was overwhelming as we watched houses reduced instantly to bits of wood, lights flashing as electric lines snapped, and cars sailing through the air. Our feelings of triumph in maneuvering around the tornado were shortlived, as the roar of a second funnel drowned out the sound of our car. Before we could react, we were sailing through the air up into the second tornado. As the tumbling car gave us an upward view we looked into a long smooth-walled tube lighted by glowing electricity. As we began to descend we saw buildings and trees zipping past at speeds that we knew would mean instant disaster if we struck them, but we were powerless to change our course. Just as we thought we were finished, we were again lifted upward. Before we knew what had happened we were deposited, as gently as we had been lifted, on top of a flat roof. As we found stairs that led us to the kitchen of a restaurant, we encountered some very surprised cooks who saw us come from the door to the roof. We told them we would be back for the car tomorrow, and left them almost as bewildered as we were. . . .

TORNADO DEVELOPMENT

Tornadoes are produced as a thunderstorm develops an organized internal structure of sufficient strength to extend the vortex from the cloud base to the ground. The severe thunderstorm that develops a tornado is normally the largest thunderstorm in a squall line or a very large isolated thunderstorm. The largest thunderstorms grow to great heights and extend from near the surface through the jetstream, where the interaction there helps provide energy for maintaining an organized internal structure that can support a tornado. This organized structure has been considered previously, but will be briefly reviewed as related to tornado generation.

The warm, moist air of the thermal updraft flows in between the double vortex structure at low levels and rises up through the storm to hit the back side and oppose the propelling winds that are pushing the storm along toward the northeast (Figure 6-1). The thunderstorm interacts with the jetstream, which provides a suction over the top of the storm, especially over the cyclonic vortex that has become better developed because of large scale cyclonic curvature within the jetstream. The cyclonic vortex then forms a link between the cloud base and upper atmosphere by providing a tubelike connection up to the jetstream level. The dynamic updraft then develops through the core of the cyclonic vortex to combine with the rotation of the cyclonic vortex to generate a vortex of sufficient strength to reach the ground. This, then, is a tornado funnel.

Many factors, therefore, are involved in arriving at a steady-state structure within a severe thunderstorm capable of tornado production. Major factors are the thermal updraft, which contains the unstable, warm, moist air; the jetstream; and the dynamic updraft through the mesocyclone or rotating southern part of the thunderstorm.

The generation of the double vortex structure occurs in the upper atmosphere in the jetstream region, and is further intensified by airflow around the north side of the thunderstorm, which adds circulation to the anticyclonic vortex, while similar flow around the southern side of the thunderstorm contributes to increased cyclonic circulation. In order to develop a tornado, the cyclonic vortex must extend down through the cloud base. The thunderstorm acts as a unit with the vortices generated in the upper part of the thunderstorm extending downward toward the cloud base where the winds at this lower altitude are moving in an opposite direction to the winds higher up.

Figure 6-1. Large tornadoes are generated by appropriate air currents within a thunderstorm. This photograph of an oil painting by the author shows the east side of a severe thunderstorm with a tornado to the south and rain area to the north. The low level inflow is opposed by the winds in midlevels of the thunderstorm creating a double vortex internal structure. The cyclonic vortex in the southern part of the thunderstorm produces tornadoes.

This has the effect of concentrating incoming air at the cloud base much as if a suction were applied in front of the thunderstorm that pulls the warm, moist air in between the vortices to form the thermal updraft (Figure 6-2). The whole thunderstorm structure is then maintained and becomes more efficient. As condensation occurs, additional heat is provided to the thermal updraft and more buoyancy is created adding further to the wind speeds. The cyclonic vortex becomes even more efficient as the circulation becomes strong enough to centrifuge the raindrops outward, thus eliminating their obstruction

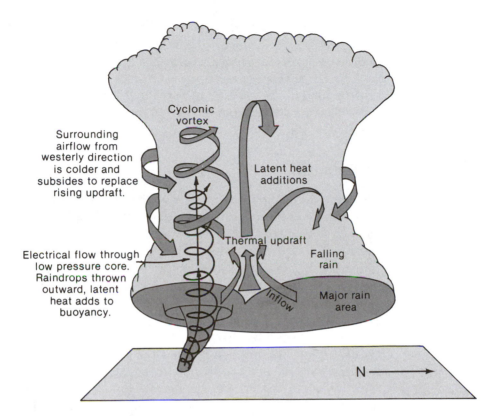

Figure 6-2. Many factors act in unison to form a tornado. The inflow of warm moist air in the front of the thunderstorm at low levels opposes the higher level westerly flow to initiate a large cyclonic vortex. Electrical flow and latent heat contribute to its strength after its formation.

effect. Lightning discharges within the mesocyclone and the thermal updraft add further to the updrafts. As the large scale atmospheric environment and the internal circulation within the thunderstorm operate in harmony, the mesocyclone becomes strong enough to support the smaller, but much more intense, circulation of its attached tornado. As we have previously seen, the high wind speeds within the tornado are possible because of the conservation of angular momentum.

Much smaller tornadoes are formed by thunderstorms that do not have the advantage of all the appropriate atmospheric conditions. Thus, the smaller thunderstorm over the Gulf of Mexico frequently generates waterspouts from the strength of the thermal updraft, combined with much smaller side winds, without the benefit of the jetstream.   Similarly, smaller tornadoes develop and last for a short time in areas where severe weather has not been forecast, as indicated by statistics that show only about 30% of tornadoes forming in forecast tornado watch areas, while about 50% of the larger, more destructive tornadoes form in forecast areas.

TORNADO DAMAGE PATHS

Tornadoes originate within the mesocyclone as the horizontal circulation and updraft core become strong enough to extend below the cloud base.   The type of tornado damage path depends on where the tornado is generated within the mesocyclone.   If it is developed in the center of the mesocyclonic rotation, and if the thunderstorm is moving at a very high rate of speed, then the tornado damage path will be long and straight (Figure 6-3).   If the tornado develops at the southern edge of the thunderstorm, which occurs quite often, the mesocyclonic circulation carries the tornado around with it in a cyclonic direction.   When the tornado reaches the front central part of the thunderstorm near the rain area it soon dissipates while another funnel forms on the southern edge of the mesocyclone.   This results in a series of tornadoes.   It is not uncommon for three or four tornadoes to be produced by thunderstorms of this type.   If the thunderstorm generates one tornado in the center of the mesocyclonic rotation, it normally lasts longer, although it may originate others on the edge of the mesocyclone as well.   If the tornado forms in the center and the thunderstorm moves at a slow rate of speed, the tornado will produce a looping path on the ground and may also last for a considerable length of time.

The three major types of damage paths of tornadoes are, therefore, the repeating type, in which a single thunderstorm produces 2 to 4 different tornadoes, one after the other; the looping type; and the straight type.   The looping damage path, like that of the Lubbock tornado, occurs from a tornado in the central part of the mesocyclone with a thunderstorm that is moving quite slowly due to light upper level winds.   The tornado takes a looping path because of the uneven nature of the earth's surface and the difficulty of taking air into the tornado core from all sides equally.   If the tornado originates in the central part of the mesocyclone and the thunderstorm is moving at a fairly rapid speed, it produces a straight path. The repeating damage paths are produced as tornadoes develop on the southern edge of the thunderstorm within the cyclonic circulation of the thunderstorm, which carries the tornado around to the front part of the storm and

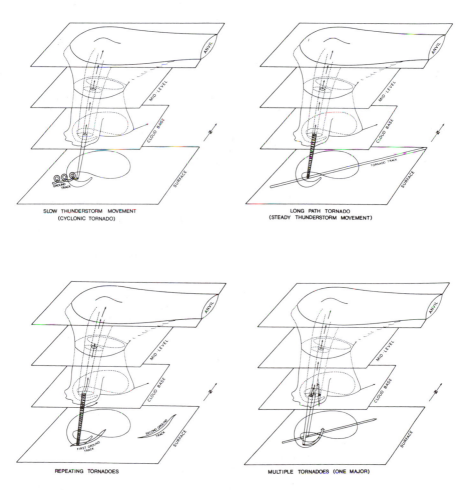

Figure 6-3. Tornado damage paths are related to the speed of travel of thunderstorms and to their location within the mesocyclonic rotation. Long path tornadoes are more centered within the mesocyclonic rotation in the southern part of thunderstorms, while intermittent tornadoes are off-centered with respect to the mesocyclonic circulation. (From Eagleman, Muirhead, and Willems, *Thunderstorms, Tornadoes, and Building Damage*, D.C. Heath & Co., Lexington, Mass. 1975.)

then into the middle where it dissipates. As it dissipates, the same conditions that caused the first one create another.

An analysis of Missouri tornadoes by Grant Darkow showed that about one-third of the tornadoes were produced by parent clouds that developed more than one tornado. These tornadoes did about half the total damage because the path lengths were a little longer than normal. An analysis performed to see if a constant interval occurred between the individual tornadoes gave some indication of a repeated time interval. Most thunderstorms that formed more than one tornado developed the second one between 30 and 60 minutes after the formation of the first. Very few developed a second tornado after 90 minutes had elapsed since the first tornado. The most common interval between tornadoes was 45 minutes with three or four typically generated from one parent cloud.

Tornadoes may stay on the ground for several hundred kilometers. In 1927, a tornado in Kansas was on the ground for 170 km. The Tri-State Tornado, in 1925, developed in Missouri, and went through Illinois and Indiana, staying on the ground for over 400 km.

The typical tornado damage path is only about 10 km in length and 150 m in width. The typical speed of a tornado is 50 km/h. The fastest ones travel at speeds up to 175 km/h. The tornadoes in the major outbreak on April 3, 1974, traveled at speeds of 100 km/h because of the strong upper atmosphere winds. If the tornado damage is 10 km long and the tornado travels at a speed of 50 km/h, it is on the ground only 12 minutes. If the parent cloud produces one tornado after another, the total life span may extend two or three hours.

PRESSURE AND WIND SPEED

Many of the characteristics of tornadoes are still largely unknown. For example, the minimum pressure within a tornado is not well known in spite of its importance. Tornadoes have passed over weather stations where barometers were in operation. However, the response time of barometers, as well as the time required for the equalization of pressure inside and outside the buildings, prevents the measurements from being very meaningful. The pressure inside a tornado is low enough to cause buildings to explode as the pressure suddenly decreases when the tornado arrives. We can only guess at the minimum pressure within a tornado. It may be as low as 60% of the normal pressure or about 600 mb.

Similarly, we do not have measurements of the wind speed within tornadoes. As the tornado passed over the weather station at Topeka a piece of debris hit the anemometer before it reached 150 km/h. On another occasion in Springfield, Missouri, a tornado that passed over a weather station spun the three-cup anomometer so fast that one of the cups blew off before it reached 150 km/h. The best estimates are obtained by looking at tornado damaged structures and estimating the amount of force required to bend particular strips of metal, for example. The light poles at Texas Tech University in Lubbock, were designed to withstand wind speeds of 250 km/h, but the tornado in 1970 toppled them over. Wind speeds great enough to force straws into poles, pluck feathers off chickens and do other strange things, commonly occur in tornadoes (Figure 6-4). An analysis of structural damage in the Dallas tornado of 1957 indicated that winds in excess of 500 km/h probably occurred. Wind speeds in tornadoes probably occur over a considerable range from 200 km/h up to the speed of sound. Some estimates indicate that as many as 50% of tornadoes have windspeeds less than 400 km/h. This makes it possible to consider designing houses and other structures to withstand winds of this speed.

APPEARANCE

A tornado is visible because of dust and debris picked up from the ground and from the condensed water vapor. Water vapor inside the funnel condenses because the pressure is less. The lower pressure inside the funnel causes the air to expand and cool with the result that water droplets condense. Figure 6-5 shows a funnel that has dust and debris in the lower part of the funnel and condensed water vapor in the upper part to make it visible. The funnel may vary from massive (Figure 6-6) to small or multiple tornadoes (Figure 6-7).

A few eye witness reports of close encounters with tornadoes are available. Milton Tabor was a student in Lincoln, Nebraska, on March 23, 1913 when a tornado originated right over his head. He describes it in *Weatherwise,* April, 1949:

The tornado cloud formed while we were enjoying a picnic and groaned furiously high in the air straight over our head. We looked up into what appeared to be an enormous hollow cylinder bright inside with lightning flashes and black as night all around. The noise was like ten million bees, with a roar that defies description.

Figure 6-4. Extreme wind speeds in tornadoes generate some unusual results. Wood is a softer material than glass but the winds of the Lubbock tornado carried the splinter shown above at such speed that it was able to penetrate this windshield even though the splinter struck the windshield at a slanted angle.

Figure 6-5. Tornadoes become visible as water vapor condenses in the vortex or as material is picked up from the ground. The hollow core in the upper part of this tornado that formed near Denver, Colorado, on May 18, 1975, is made visible by water droplets. (Courtesy of NOAA.)

DISTRIBUTION AND NUMBER

Tornadoes are most common in the month of May, but more days in June (26 out of 30) have tornadoes than any other month (Figure 6-8). In general, the winter months have few tornadoes while the spring months from April through June have many more. The number of tornadoes that occurs each year averages 120 in Texas alone, while Oklahoma has 58, and Kansas is third with 48 tornadoes. If the number of tornadoes per unit area is used, Oklahoma becomes the greatest tornado producer. These results are based on statistics gathered since 1956; prior to this time the records are not as accurate since the number of tornadoes per year increased because of improved reporting. Kansas, in spite of Wizard of Oz fame, is third in both

Figure 6-6. The larger circulation surrounding the tornado is evident in this photograph taken near Union City, Oklahoma. (Courtesy of NOAA.)

total number of tornadoes per year and number per unit area. Surprisingly, Florida is fourth in both categories followed by Iowa and Missouri, on a unit area basis.

More tornadoes occur in the Central United States than anywhere else in the world because of the unique combination of weather conditions and topography. The mountains to the west of the Great Plains cause the air flowing over them to subside as it flows into the Great Plains. This creates upper air inversions that affect cumulus cloud development. Surface weather patterns frequently include a high pressure area off the East Coast of the United States in the Atlantic Ocean. This brings south winds from the Gulf into the Great Plains area (Figure 6-9). The lower atmosphere frequently has very hot, humid air from the south winds at the surface with cooler, drier airflow above from the west as the air flows over the mountains. This unique combination sets up the proper atmospheric conditions for severe thunderstorms and tornado generation as previously described.

The number of deaths from tornadoes in different states does not match the total number of tornadoes. Tornadoes cause more deaths in southern states, such as Mississippi and Alabama, than in other locations (Figure 6-10). Kentucky also has a high death rate per year. Midwestern states that have more tornadoes have fewer deaths; Kan-

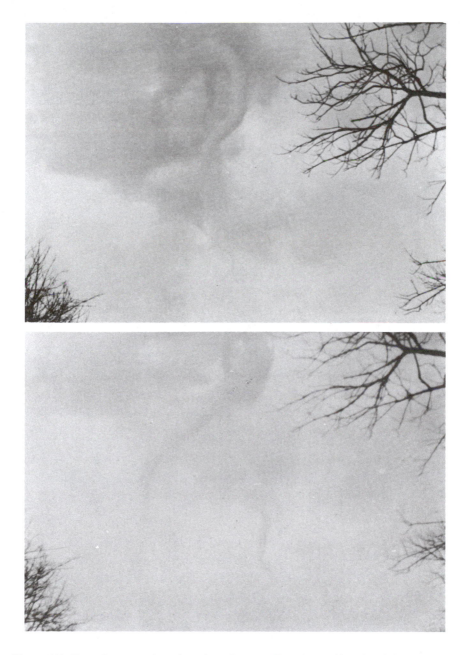

Figure 6-7. Tornadoes come in various sizes, shapes, and numbers. This series of photographs shows multiple funnels from a single cloud. (Courtesy of James D. Tatum.)

sas and Oklahoma, for example, average only seven deaths annually in spite of the greater number of tornadoes there. This indicates that preparation and warnings are important in protecting people from the storms.

The years 1973 and 1974 were memorable for tornadoes. Although the average number of tornadoes per year is 700, a total of 1109 tor-

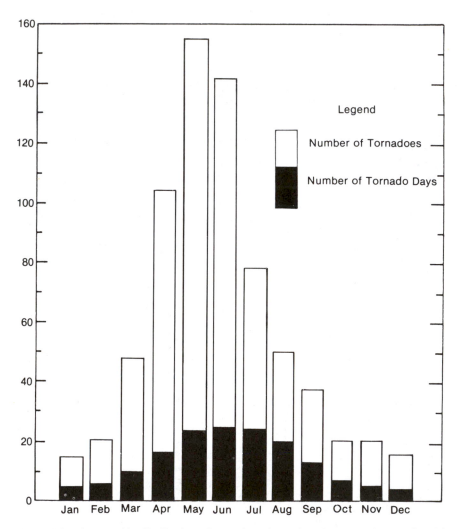

Figure 6-8. The monthly distribution of tornadoes shows that more occur in May than any other month, but more days in June have tornadoes than other months. This distribution is based on 18,641 tornadoes that occurred from 1953 to 1978. (NOAA data.)

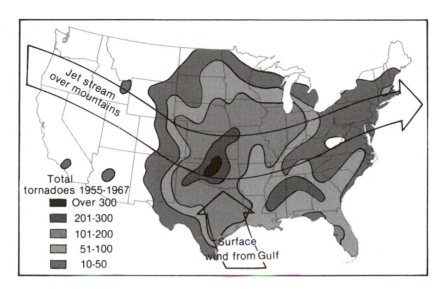

Figure 6-9. The central United States is the favored location of tornadoes as the Rocky Mountains block surface winds. The jetstream flows over them while warm surface air is fed into the Central Great Plains. This causes more tornadoes to form in a band extending from Texas through Kansas and into Missouri. However, each of the fifty states have experienced the effects of tornadoes.

nadoes occurred in 1973, to break the previous record for the most tornadoes in any year, set in 1967 with 929 tornadoes. A new record for the total number of days in the year with tornadoes was also set in 1973 when 208 out of 365 days had tornadoes. Tornado records of a different type were established in 1974. A record number of tornadoes, fatalities, and injuries occurred during a particular outbreak on April 3 and 4. Although the total number of tornadoes for the year was quite high (945), 144 of these occurred within a two day period, April 3 and 4. A total of 361 people were killed by tornadoes in 1974 and 6,915 more were injured. People were killed in 20 different states by 64 different tornadoes. Alabama had more deaths than any other state with 79 fatalities, but the single largest killer tornado occurred at Xenia, Ohio where 34 people were killed on April 3. Most of the deaths in Alabama (77) occurred during the April 3 and 4 outbreak. Thirteen different states had major tornadoes during this outbreak. Eleven of these states suffered fatalities as a result of the tornadoes during this period. According to two different accounts, 144 or 148 tornadoes were sighted; some of these crossed state

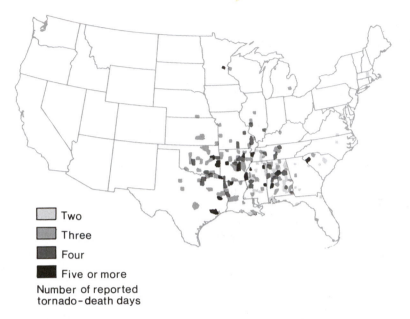

Two

Three

Four

Five or more

Number of reported
tornado-death days

Figure 6-10. Number of days with tornado deaths, by county, from 1916 to 1953 shows that more tornado deaths occur in the southern United States. This distribution does not match either the number of tornadoes or the concentration of population. (From Tornado Deaths in the United States, U.S. Linehan, US Government Printing Office, 1957.)

lines to account for some records showing as many as 153 tornadoes. This super outbreak of tornadoes on April 3 and 4 caused 307 deaths and 5,514 injuries with over half a billion dollars damage to property.

TORNADO DETECTION

It is the responsibility of the National Severe Storms Forecast Center in Kansas City, Missouri to issue tornado watches based on such atmospheric factors as the jetstream, upper air divergence, presence of a squall line, and other factors as previously described. Such tornado watches are issued for large geographic areas covering several thousand square kilometers. Tornadoes, of course, only affect a very small portion of this area.

A tornado warning indicates that a direct threat of tornadoes exists in a localized area. Warnings are the responsibility of the local weather service, Civil Defense Director, and television or radio station operators. They are issued on the basis of the observation of a hook echo on the radar screen or from eyewitness reports, normally verified by highway patrolmen. If the local weather station is equipped with radar, the formation of a hook shaped echo is the usual basis for issuing a tornado warning. This warning procedure is not as effective as generally believed by most people. The hook shape on the screen only indicates that the thunderstorm has generated a mesocyclone which may be capable of producing a tornado. More than half of the tornadoes occur from thunderstorms that do not produce a hook shaped echo on the radar screen. In order to see the hook the radar must be aimed at the lower one-fourth of the thunderstorm. Thus, the range of the radar set is limited for severe thunderstorm detection because of the earth's curvature.

The hook echo (Figure 6-11) forms as the radar displays the reflection from raindrops within a horizontal slice through the lower part of a severe thunderstorm. The radar signal is reflected back to the station by raindrops carried in the outer part of the mesocyclone. This circulation brings rain originating in the thermal updraft around to the front part of the thunderstorm forming the hook shape on the

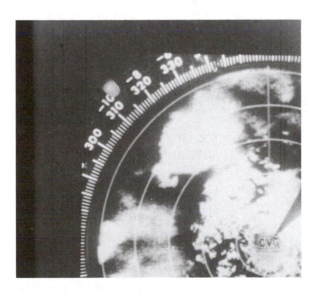

Figure 6-11. Hook shaped echoes on a radar screen are one of the means of detecting severe thunderstorms and possible tornadoes. The hook shape is caused by the cyclonic rotation in the southwestern part of a thunderstorm.

radar screen. Most thunderstorms that form a hook echo also develop a tornado, although the tornado does not show up on a radar screen.

Storm spotters are an important part of the warning process in some communities. Lawrence, Kansas, for example, has an organization of 50 storm spotters who use ham radios and CB radios to report to the Civil Defense Director. The spotters are assigned to a particular location from which they report on developing severe weather during a tornado watch. Johnson County, Kansas has one of the best spotter organizations in the nation. The Johnson County operation is coordinated by the Civil Defense Director who receives reports from spotters in various preplanned locations. The Civil Defense office contains a communications system with a line to the Severe Storms Forecast Center and various TV and radio stations so that information can be passed quickly and efficiently. Dozens of tornado sirens are distributed throughout the county to be sounded when tornadoes are approaching. Some public places also have specially designed radios where the Civil Defense Director can announce developments over the appropriate frequency to reach a large number of people in a hurry. Even with this elaborate warning organization, available technology is not utilized to its fullest. The tornado sirens (Figure 6-12) that are operated as a warning device do not cover the area very effectively. The sound from sirens is not likely to penetrate to people inside houses during severe weather with lightning, thunder, and high winds. Thus, the use of the specially designed radios is a trend of the future. The National Weather Service also broadcasts weather information continuously at frequencies of 162.4 to 162.55 MHz from a large number of stations over the United States.

Individual warnings may be provided by the noise of a tornado. The noise has been variously described as ten million bees, or many freight trains or jet airplanes. It is basically a noise so loud that it may provide a warning in itself. The source of the sound is uncertain, but it may be the result of supersonic winds within the thunderstorm.

The appearance of the sky may indicate tornado formation. A "tornado sky" is shown in Figure 6-13. Such clouds form when the atmosphere is very unstable and are usually present with tornadoes although their formation does not always mean that a tornado has also formed.

Animals may also provide tornado warnings to people. Dogs, for example, hear sounds at ranges beyond human capabilities (the prin-

Figure 6-12. Many cities have installed tornado sirens as a part of a public warning network. Such sirens are normally sounded by the Civil Defense Director in cooperation with the National Weather Service.

ciple behind silent dog whistles). There are indications that animals can also hear tornadoes at greater distances than people. Dogs may drastically alter their behavior when a tornado is approaching. A dog that never comes inside the house may scratch on the door and come in when the door is opened and crawl under a bed as a tornado approaches.

Your television set may be used for tornado warnings. If you turn the selector to Channel 13, darken the screen, and switch to Channel 2, the presence of a glowing white screen indicates the presence of a tornado within 40 km. Channel 2 operates at a frequency that is

Figure 6-13. A tornado sky such as shown here indicates severe turbulence in thunderstorms and usually is present when tornadoes occur. Some thunderstorms, however, have this mammatus formation and never form a tornado.

close to that of the electromagnetic radiation coming from severe thunderstorms that contain tornadoes. The method may not be entirely accurate since all severe thunderstorms apparently do not produce enough radiation to be detected by Channel 2.

A commercial tornado detector is also manufactured although it is still experimental (Figure 6-14). It consists of a radio that operates at 3 MHz with an antenna that picks up the electromagnetic radiation from severe thunderstorms. The amount of detected radiation is displayed by flashing red lights and a buzzer that sounds if the burst rate of energy from the storm is great enough.

Radios can be purchased that pick up one of the few hundred National Weather Service radio stations where continuous weather information is broadcast at a frequency in the public service band at about 162.5 MHz. These stations also send a signal that automatically turns on radios manufactured for this purpose. This signal is broadcast when tornado warnings are issued. Such specially equipped radios may prove to be an important warning device in homes, schools, and public places.

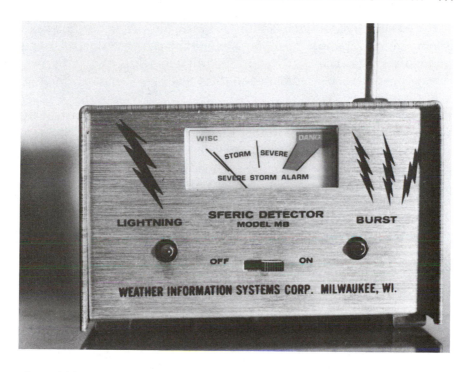

Figure 6-14. This commercially available detector measures the electrical activity within severe thunderstorms. Although some tests indicate a high false alarm rate, experimental instruments such as this may be of value in detecting severe thunderstorms.

TORNADO SAFETY

The chances for surviving a major tornado can be improved greatly by advance preparation and proper response during a tornado warning. Advance preparation will be considered in a later chapter dealing with building design and construction techniques. A person's immediate response to a tornado threat should be governed by a good understanding of tornado safety factors.

The safest location in a tornado stricken area is within underground concrete shelters. Rooms or storm shelters constructed of at least 25 cm of reinforced concrete for the walls and ceiling are ideal shelters. Basements offer considerable protection from tornadoes if they are constructed of reinforced concrete. Damage surveys have shown that safest locations are opposite the approach direction of the tornado; thus the northeast rooms are safer.

Safest locations in houses without basements are in the northeast rooms (rooms opposite the approaching tornado) with smaller rooms and closets safer than large rooms. Boards and debris frequently penetrate the south and west walls and these walls are also more likely to fall inward than others.

Automobiles offer little protection from a tornado if they are caught in it. Many of the deaths during the Wichita Falls tornado occurred in automobiles. A large tornado traveling at 100 km/h is not as easy to maneuver around as might be expected, especially if traffic is heavy.

Mobile homes are not a safe location during tornadoes. Mobile home parks should provide a community shelter, perhaps an underground recreational area, for use by residents.

If you are caught in open country during an approaching tornado seek a small depression such as a ditch running perpendicular to the path of the tornado. This may be helpful since winds and debris traveling at speeds of several hundred km/h will not make sharp bends.

SUMMARY

Tornadoes touch down as the circulation within the thunderstorm develops sufficiently to support them. The mesocyclone increases in strength as the surrounding airflow passes around the southern side of the thunderstorm, lightning heats the inner walls of the vortex, and rain is thrown outward by the centrifugal force of the circulation.

Tornadoes that develop in the center of the mesocyclone last longer and produce straight damage paths from faster moving thunderstorms and looping damage paths from slow moving thunderstorms. Tornadoes on the southern edge of the mesocyclone are carried by the thunderstorm rotation around to the front of the thunderstorm and then into the rain area where they dissipate. Such a pattern is frequently repeated with the formation of several different funnels.

The typical tornado damage path is 150 m wide and 10 km long, with a forward speed of travel of 50 km/h. Accurate measurements are not available to tell us the amount of pressure decrease in the tornado funnel or the speed of the winds.

Tornadoes are more frequent in the Central United States, but more people are killed by tornadoes in the Southern United States. A single tornado outbreak in 1974 killed 307 people.

Tornado watches are issued by the National Severe Storms Forecast Center when the atmospheric conditions are right for tornado development. Tornado warnings are issued when a direct threat of tornadoes exists. Warnings are based on the observation of a hook echo on the radar screen or eyewitness reports of tornadoes on the ground, and are issued by local Weather Service offices or Civil Defense directors, in cooperation with radio and television stations. Individual warnings may be provided by the television technique or special weather radios.

The safest location in a tornado stricken area is in an underground reinforced concrete room or shelter. Concrete basements are safer than houses without basements. Northeast rooms in the basement and first floor are safer than other rooms for a typical tornado that approaches from the southwest.

7
Fire From Above

At least we all made it to the 18th hole green before the rain. The rumbling from the darkened western sky had only vaguely registered as I tried to concentrate on the previous couple of shots. It was obviously going to rain as the growing thunderstorm changed the complexion of the afternoon sky drastically from its previous sunny state. We had been wet before while finishing a round of golf, and the extremely hot afternoon made the prospect almost inviting.

No one rushed the last round as the leading edge of the thunderstorm moved over us. We had heard thunder in the distance, but had no real sense of danger until the whole atmosphere seemed to be charged. My hair felt filled with electricity, and a glance at the others revealed the astonishing fact that their hair was literally standing on end. Just as I seemed to recall some advice for this situation the lightning discharge struck so close that I wondered if I was dead. When we recovered we saw that the flag on the rod in the cup only 4 steps away was burned to a crisp by the lightning discharge. Lucky for us that we hadn't removed it yet, and that it was taller than us by the length of a hair. . . .

HISTORICAL SETTING

Even small cumulonimbi are capable of generating thunder and lightning, giving rise to their common name of thunderstorms. The largest thunderstorms that produce tornadoes, rain, and hail also produce almost continuous lightning. Throughout history man has always had respect, awe, and even worship for lightning. He has not, however, always known how to deal with it. For example, in the eighteenth century it was the practice in Europe to ring church bells when thunderstorms developed, with the hope that this would disrupt the continuity of lightning paths and drive the lightning away. This practice was so dangerous to bell ringers that the people of Paris passed a law in 1786 making bell ringing during thunderstorms illegal.

The problem of living with the lightning hazard in the eighteenth century was amplified because churches were used for storing gunpowder. Most of the churches were designed with tall towers making them attractive targets for lightning discharges. Lightning and exploding gunpowder in churches were responsible for killing thousands of people in the eighteenth century. Since the nature of lightning was not understood at that time there was no information on where lightning would strike or why it struck particular locations.

It is interesting to note that some tall buildings were never touched by lightning even though they were constructed long before lightning was known to be caused by thunderstorms and electrical charges. One of these was the temple built by Solomon in Jerusalem. According to historians it experienced no damage from lightning over a ten century period. The wood and stone structure was completely covered with a thin layer of gold and had metal spikes on top of the temple protecting it from lightning.

STUDIES BY BENJAMIN FRANKLIN

The first studies of electricity were conducted around 1600 by William Gilbert, a physician for Queen Elizabeth I. He performed experiments that showed that material such as amber and glass would attract light bodies when rubbed. He gave this the name electrics taken from the Greek word electron, meaning amber. More than a century passed before much more progress in understanding electricity was made.

In 1746, Benjamin Franklin, at the age of 40 following successful business ventures, became interested in electricity. He discovered positive and negative charges and invented an electrical machine that could be cranked to generate sparks several inches long. He designed other experiments to gain more information on atmospheric electricity since very little was known about it at that time. He devised an experiment to confirm that electricity was developed during thunderstorms. The experiment required a tall tower that could be insulated from the ground near its base. Franklin suggested that a spark could be produced between a gap between the insulated tower and the ground by bringing a grounded wire near the metal base of the insulated tower (Figure 7-1). Although he had no tall tower in Philadelphia for testing it, he publicized this idea in Poor Richard's Almanac that

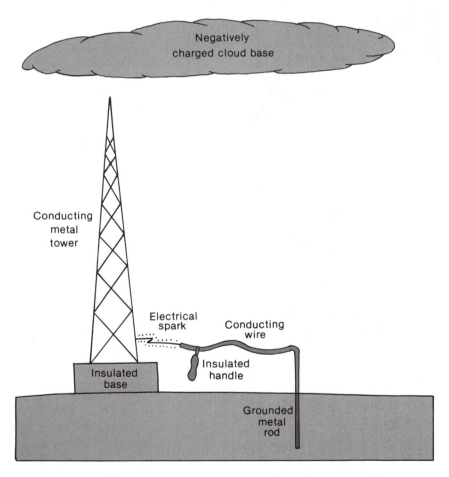

Figure 7-1. Experiment designed by Benjamin Franklin to test the charge buildup within thunderstorms. He suggested that a conducting wire be brought close to the insulated base of a tower while thunderstorm activity occurred overhead.

was selling at the rate of 10,000 copies a year. The experiment was successfully demonstrated in Paris, France in 1752 when a spark was produced every time a thunderstorm came over the tower used for the experiment. Benjamin Franklin became famous in France and even before he knew the experiment had been successfully conducted he was known as the man who could bring fire down from the heavens.

In the meantime, Benjamin Franklin was performing experiments with a kite. He was not satisfied just to know that thunderstorms contained electricity, but he tried to discover whether it was positively or negatively charged. In the spring of 1753, he found that all thunderstorms were negatively charged, but on June 6 of that year he measured positive charges during a thunderstorm. He concluded that nearly all thunderstorms have negatively charged bases, but occasionally one is positively charged. This remained the only reliable information on electricity in thunderstorms for 170 years. It wasn't until the 1920s that additional electrical experiments were performed to provide more information on the electrical nature of thunderstorms.

Benjamin Franklin invented the lightning rod with the first ones installed in 1756. A few years later in 1760, lightning struck a house in Philadelphia that was protected by a lightning rod and received no damage. Franklin had installed a lightning rod on his house, but it was not struck by lightning until 1787 when he was 81 years old.

Benjamin Franklin's understanding of electricity was extraordinary for his time. He amused his friends by giving them electrical shocks during his kite experiments. Others such as King Louis XVI used a kite to shock 200 monks who were all holding hands. The electricity flowing through the kite string was also dangerous since Professor Richmann in St. Petersburg was killed by ball lightning while performing experiments with a kite during a thunderstorm.

FORMS OF LIGHTNING

Most thunderstorms become negatively charged at their base and positively charged near the top through processes to be described later. The most common form of lightning (over 60%) is intracloud lightning, as discharges within a cloud occur between regions of positive and negative charges.

The next most common type of lightning is forked, or cloud to ground lightning. The lightning discharge is between the cloud base and the ground (Figure 7-2). The discharge develops as a negatively charged cloud base induces a positive charge on the surface of the earth by repelling other negative charges. Since opposite charges attract and similar charges repel each other, the negative charges at the thunderstorm base repel the negative charges within the surface layer of the earth. Therefore, the positive charge on the earth during a thunderstorm is an induced charge arising from the charge generation within the thunderstorm. Forked lightning forms when the charge separation between the earth's surface and the base of the cloud develops to a critical level, just as electricity sparks across the gap in a spark plug, or from your finger to the door knob after walking over a nylon carpet.

Fig. 7-2. A common form of lightning is forked lightning that occurs as a discharge develops between the cloud base and the earth's surface.

There are normally several individual lightning strokes during a single lightning flash. A flash of lightning lasts about 0.5 sec and is typically composed of three separate strokes, although as many as 25 strokes have been observed during a single lightning flash. Such a discharge lasts for more than a second and is, therefore, much more visible than the typical lightning flash.

Ribbon lightning sometimes forms if the winds are strong enough to displace the conducting channel beneath the cloud during the separate lightning strokes. A wind speed of only 30 km/h would displace the channel 8 m in one second. During a half second lightning discharge the multiple strokes through the same channel would be separated by a distance of 4 m between the location of the path of the first and last stroke. The diameter of the lightning discharge channel is only a few centimeters. This channel conducts the charge from the cloud to the earth and from the earth back to the cloud during a typical lightning discharge. If winds displace the channel enough for it to be visible, the result is ribbon lightning.

Sheet lightning is the type that occurs from distant thunderstorms as a flash beyond the horizon lights up the sky. Heat lightning is similar to sheet lightning except that it occurs from thunderstorms located even further away so that no thunder is heard and the flashes are fainter.

Another form of lightning is an air discharge (Figure 7-3). This is a discharge into the air beneath the thunderstorm as the leader stroke (to be explained later) fails to reach the ground. This occurs most frequently in desert regions where the lifting condensation level is higher in the atmosphere.

A bolt from the blue is a discharge that begins as an air discharge, but after traveling a few kilometers, one of the leaders reaches the ground producing a lightning strike (Figure 7-4). The point that is struck may be as many as 15 kilometers away from the thunderstorm. The bolt from the blue is, therefore, produced by a thunderstorm that is a little farther away than normal.

St. Elmo's Fire is a form of electricity that is most likely to be seen at the peak of tall objects. These may have a bright glow because of the positive charges that remain as the negative charges drain away when a thunderstorm moves overhead (Figure 7-5). Towers, mountain peaks and other tall objects, even the horns of cattle, sometimes display this kind of electricity. Occasionally, the top of a thunderstorm

Figure 7-3. Air discharges occur when lightning is unable to complete a path from the negatively charged cloud base to the ground. Air discharges are more common in dryer regions where the cloud base is higher.

glows with something like St. Elmo's Fire as the positive charges are concentrated there at extremely high levels.

Pilots of aircraft frequently observe St. Elmo's Fire if they must fly through thunderstorms. All commercial aircraft have radar on board to avoid thunderstorms, but this is not always possible. The lightning that sometimes strikes aircraft may be preceded by St. Elmo's Fire by five minutes or more. If lightning strikes an airplane it normally passes along the greatest distance, nose to tail or wingtip to wingtip, since the airplane is simply serving as a connection be-

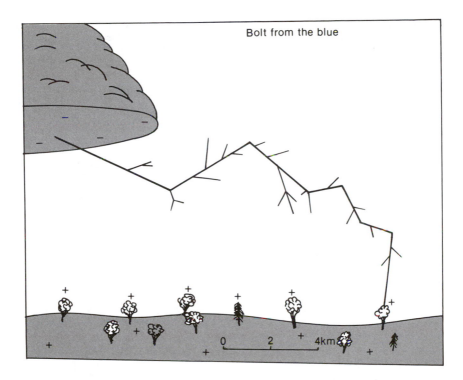

Figure 7-4. A bolt from the blue occurs as the lightning travels for some distance horizontally before completing a path down to the ground. Such lightning may occur as far as 15 km from the thunderstorm.

tween the different parts of the cloud. If the plane is all metal it is a better conductor than the air for transferring the charge from one location to another. Metal aircraft are not usually damaged by a lightning strike. However, wood or fabric located between conducting metal may be ignited if struck by lightning.

Ball lightning is a form of lightning that is not well understood. An investigation of several hundred reports of ball lightning indicated that only three were possibly actual fire balls. The others could be accounted for by such explanations as images left on the eye after an intense lightning flash, since ball lightning usually occurs near ordinary lightning flashes.

Ball lightning consists of a ball of fire from 1 cm to 2 m in diameter. These may fall from the sky and explode or roll downhill until they strike an object and explode. They have an appearance resem-

Figure 7-5. St. Elmo's Fire develops as the positively induced charge on tall objects on the earth develops sufficiently to cause an air glow. St. Elmo's Fire may precede a lightning discharge although the lightning does not always follow.

bling a soap bubble or a bubble of electricity. Ball lightning has come inside houses through windows or electrical outlets to float or roll across a room. One ball of fire about half a meter in diameter was reported to roll down a hall toward a person who stepped aside to let it pass.

When ordinary lightning strikes in deserts it leaves permanent evidence. The lightning flash melts some of the sand and forms a glassy channel a few centimeters in diameter and a meter or more in length. These glassy ribbons of glass, called fulgarites, have been used to study the climate of dry areas such as the Sahara Desert. The number and location of fulgarites is related to the number of thunderstorms and therefore gives some estimate of thunderstorm activity in the past.

DISTRIBUTION OF THUNDERSTORMS

Since lightning is developed by processes within thunderstorms the distribution of thunderstorms gives some information on the occurrence of lightning. In the United States, the average number of thunderstorms is greatest in the Florida Gulf Coast area where the mean number of days with thunderstorms is about 80 per year as previously shown in Figure 4-1. Another area that has a high frequency of thunderstorms is Colorado and New Mexico with thunderstorms on about 60 days during each year. The fewest number of thunderstorms occurs along the West Coast. In spite of the fact that the Pacific Northwest has the greatest mean annual rainfall in the United States, very few thunderstorms occur there. The rain in Washington and Oregon falls more as a continual mist.

Over the whole world about 44,000 thunderstorms are in action every day, while about 2,000 thunderstorms exist at any one time. These produce an average of about 100 lightning flashes per second. Even in the central Sahara Desert one thunderstorm per year normally occurs.

ORIGIN OF THUNDER

The thunder associated with a thunderstorm is produced by the discharge of electricity. Much of the energy of the lightning discharge is used in heating a channel of air to conduct the electricity from the thunderstorm to the ground. The conducting channel is heated to $30,000°K$, about five times the temperature of the surface of the sun. As the air in this channel is heated very rapidly, in a few millionths of a second, it suddenly expands from a few millimeters to a few centimeters in diameter. This central conducting core is surrounded by a glow discharge that may be several meters in diameter. The shock waves that result from the sudden expansion of air in the conducting channel spread outward as thunder.

The speed of sound is much slower than the speed of light. The sound wave travels only one kilometer in three seconds. This can be used to estimate the approximate distance to the lightning discharge by measuring the time in seconds between the lightning flash and the sound of thunder, and dividing by 3 to obtain the distance in kilometers. The sound waves travel for only about 10 km, so thunder will not be heard past this distance.

Rumbling occurs as discharges coming from different parts of the cloud arrive in short succession, with sounds also altered by echoes between clouds to produce a rumbling sound instead of the sharp crack produced by the main trunk of a nearby lightning discharge (Figure 7-6). The branches of the expanding air channel provide the crackling sound that accompanies many lightning flashes.

DEVELOPMENT OF ELECTRICITY IN CLOUDS

A thunderstorm is able to generate lightning in about 20 minutes. Therefore the charge separation mechanism within the thunderstorm

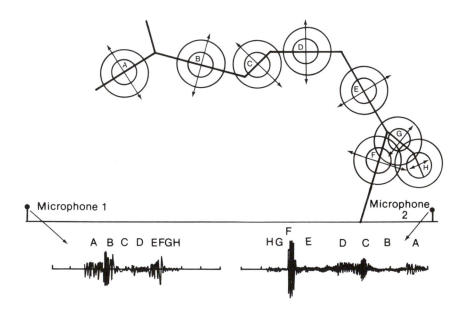

Figure 7-6. Sound waves from the main lightning channel as well as the branches combine to produce the overall pattern of claps and rumbles produced by a lightning discharge. The amplitude and duration of the sound wave produced by each of the individual elements is determined by the orientation of the segment and its distance from the observer. (After *Scientific American*, Vol. 233; 1, 1975.)

must be rapid enough to operate within this time limit. A part of the explanation of charge separation involves the sudden freezing of liquid water droplets. Large updrafts can cause rapid freezing just above the freezing level in thunderstorms. It has been observed in the laboratory that a charge separation occurs if a temperature gradient exists in ice (Figure 7-7). Some of the molecules within ice are dissociated into positive and negative ions. Warmer temperatures cause greater dissociation. If a temperature gradient exists across ice it will result in the coldest part of the ice taking on a positive charge since positive ions are free to migrate from regions of high to low concentration while negative ions are not. This is called the thermoelectric effect.

As a supercooled water droplet freezes from the outside a temperature gradient may arise due to colder external temperatures. Small splinters of ice from the outer portion of the ice crystal are positively

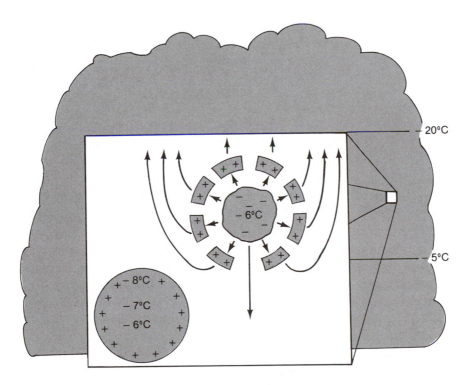

Figure 7-7. The charge separation in thunderstorms is related to the shattering of ice crystals as they are carried higher in the thunderstorm where the temperature is colder. (After Eagleman, *Meteorology: The Atmosphere in Action*, D. Van Nostrand Co., 1980.)

charged from ion migration, and are carried upward by updrafts to accumulate in the upper regions of the cloud, while the larger negatively charged ice particles fall to the lower part of the cloud. The region where this mechanism is effective is just above the freezing level, from $-5°C$ to $-30°C$.

It has been well established that large updrafts and downdrafts exist in active thunderstorms. (This is another reason thunderstorms are avoided by pilots, in addition to the lightning and hail hazard.) The large updrafts further contribute to charge separation and the development of lightning within the required time of 20 minutes. After a slight buildup of negative charge at the base and positive charge at the top of a cloud by the shattering of freezing cloud droplets, the large ice pellets have an induced negative charge on their tops and positive charge on their bottoms. When this occurs the smaller cloud droplets within an updraft strike the lower surface of large ice crystals or pellets and bleed off some of their positive charge, leaving the small droplets positively charged and the large ice pellets negatively charged (Figure 7-8). The smaller droplets are then carried to the top of the thunderstorm by updrafts, producing a positive charge there, while the larger ice pellets with a more negative charge fall into the lower part of the cloud, giving it a negative charge. Thus, the small cumulus cloud may grow into a tremendous electricity generator capable of developing almost continuous lightning discharges from the only resources available—air, water and dust particles.

It has been observed that a relationship exists between lightning discharges and rain on the ground. A major lightning flash often appears to be followed by a downrush of large raindrops at the ground. The electrical charges may help hold the cloud together as the positively charged cloud top pulls on all the negatively charged water drops at the base of the cloud. A sudden release of the electrical charges by lightning may allow the raindrops to fall more rapidly.

NATURE OF THE LIGHTNING FLASH

The lightning flash consists of several strokes. The first of these is a slower, stepped leader stroke. The stepped leader is slower because it must develop the conducting channel from the thunderstorm base to the ground. As the charge builds in the cloud and the charge separation reaches the sparking level it seeks the best channel to the high-

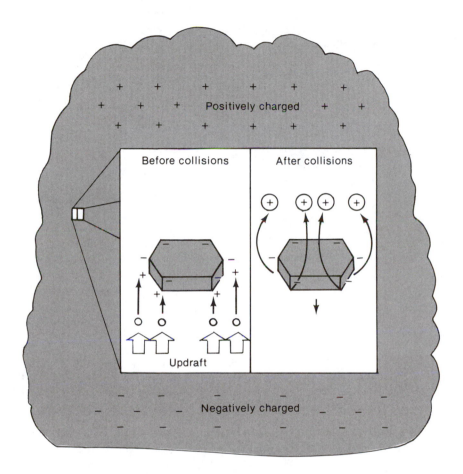

Figure 7-8. After a charge separation is developed in a thunderstorm because of the thermo-electric effect, ice crystals within the central part of a thunderstorm have a positive charge on their lower surface because of induction. There positive charges are bled off by updrafts containing smaller water droplets, thus intensifying the charge separation. (After Eagleman, *Meteorology: The Atmosphere in Action,* D. Van Nostrand Co., 1980.)

est or best conducting ground location producing forked paths since many of the channels do not lead anywhere. Eventually a path is developed from the cloud to the ground.

After the stepped leader develops a conducting channel to the ground the return streamer occurs (Figure 7-9). This is a discharge of electricity from the ground back to the cloud. The return streamer is a lot more brilliant than the leader stroke and is ordinarily the observed

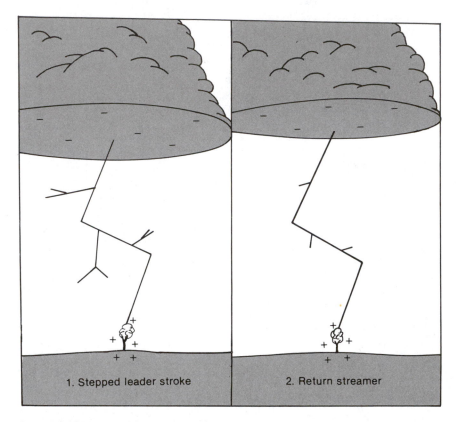

1. Stepped leader stroke

2. Return streamer

Figure 7-9. The stepped leader first builds a lightning channel down to the ground. This is followed by the return streamer as a discharge occurs back from the ground to the cloud.

lightning stroke because of this brilliance. After a channel has been developed an average of three leader strokes and three return streamers occur through one specific conducting channel. The channel is made conductive for electricity by the extremely hot temperatures that develop ionized air molecules. Oxygen molecules, for example, are ionized into single oxygen atoms. Since these are charged, the electricity readily flows through the channel.

Lightning can be photographed by an ordinary camera at night by opening the lens when a thunderstorm is within view (Figure 7-10). A lightning discharge then produces an image on the film. An ordinary camera cannot usually obtain lightning photographs during the day since an open lens will overexpose the film before lightning occurs,

Figure 7-10. Forked lightning is the most common visible form of lightning. It consists of one or more main channels from the cloud to the ground with branches out from these. (Courtesy of NCAR.)

but unusually long discharges can be photographed during the day by watching closely for them.

The successive lightning strokes within a single flash discharge layers at increasing altitudes within the cloud. The first return streamer may neutralize the lowest layer of the thunderstorm with the next streamer neutralizing a slightly higher layer and so on. This allows more of the negative charges in the thunderstorm to be neutralized than would otherwise be possible from a single stroke of lightning.

Occasionally a "superbolt" of lightning may develop that is several times stronger than an ordinary discharge. Such bolts may flow between the positively charged top of a thunderstorm and the earth, since the ground is negatively charged before a thunderstorm moves over it to induce a positive charge. Thus the charge is opposite to an ordinary lightning discharge beneath a cloud and the distance is greater as a superbolt forms (Figure 7-11).

Figure 7-11. Occasionally a superbolt lightning discharge occurs from the positively charged top of a thunderstorm to the negatively charged earth some distance away. Because of the greater distances, this lightning discharge may be 100 times as strong as an ordinary lightning discharge. (Courtesy of Alan Moller.)

One of the beneficial effects of lightning is the production of nitrogen fertilizer. As the air is heated by the leader stroke oxygen and nitrogen molecules are combined into a form usable by vegetation. This nitrogen fertilizer is carried by rain into the soil for plant use. Although the amount of fertilizer production in areas of heavy thunderstorm activity may amount to only 1 kg/hectare (5 lbs/acre) each year, it is interesting to note that it is applied at exactly the time when it is most useful to the growing plant. In comparison, 45 kg/hectare (250 lbs/acre) of man-made nitrogen fertilizer applied to a crop in dry weather is of no value and may even "burn" the crop if the dry weather continues.

LIGHTNING PROTECTION

Lightning damages amount to about $25 million per year in the United States. The best way of protecting houses from lightning is by use of lightning rods. Investigations of hundreds of fires have shown that 95% of the buildings that burned from lightning had no lightning rods and most of the remaining 5% with lightning rods had major defects in their methods of installation. The lightning rod is effective in protecting buildings from damage only if it is properly grounded.

A lightning rod has a zone of protection beneath it in a cone of about 45° (Figure 7-12). A lightning rod consists of a metal rod extending above the highest point of the building to be protected with the other end of the rod extending to moist soil beneath the surface of the ground. One meter or less beneath the surface is generally sufficient to anchor a rod into moist soil. Sharp bends in the lightning rod near other grounded routes that the lightning could take must be avoided. Lightning can jump across if sharp bends occur in the rod near metal pipes or other grounded conductors.

The spread of electricity from a lightning strike on a golf course can be seen in Figure 7-13 from the burned grass. The flag was obviously not well grounded in deep moist soil and the electricity was neutralized as it spread horizontally. This effect would be expected if only the first few centimeters of soil were moist with a dry layer beneath.

Taller objects are more likely to be struck by lightning. An object 100 m high located in a region where there are 30 days with thunderstorms per year can be expected to receive three lightning strikes per year. An object 300 m high could be expected to receive about ten strikes per year if located in an area with a similar frequency of thunderstorms. Radio, TV, and other towers are continually struck by lightning. Radio and television stations generally have auxiliary generators for this reason.

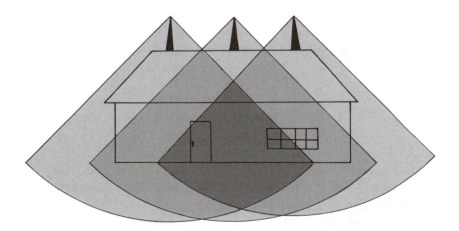

Figure 7-12. Lightning rods can be expected to protect an area enclosed within an angle 45° outward from the rod. The number of lightning rods required to protect a particular house is therefore determined by the height of the rods and the size of the house.

Figure 7-13. Lightning currents are evident in this photograph taken three days after a lightning strike on a golf course in Tucson. The 1.4 m fiberglass flag pole was tall enough to attract lightning. (Courtesy of E. Philip Krider.)

Lightning kills more people in the United States each year than any other weather event. Tornadoes cause much more property damage, but fewer fatalities than the 198 per year from lightning (Table 7-1). Lightning deaths are normally singular events, but occasionally there are multiple deaths, such as occurred in December, 1963, when lightning struck a jet passenger plane over Elkton, Maryland, killing all 81 persons aboard. The second deadliest lightning strike occurred in July, 1961, near Clinton, North Carolina, when eight of nine people were killed as lightning struck a tobacco barn and passed through the metal heating system that the victims were leaning against. Another lightning tragedy occurred in August, 1960, when a young couple in Bay City, Michigan, were struck and killed by separate bolts of lightning while dashing across the street to their home. Lightning deaths are slightly more common in the eastern and central United States than in the western United States (Figure 7-14).

Lightning deaths occur more frequently to persons in open country than inside buildings. About 52% of the fatalities occur in the open, 38% in houses or barns, and 10% under trees (Figure 7-15). Unsafe locations inside houses are near metal water pipes, metal appliances, and telephones. Dangerous locations outside are on mountaintops

TABLE 7-1. Annual Storm Fatalities in the United States
from 1940 to 1975*

	Deaths			
	Lightning	Tornado	Flood	Hurricane
1940	340	65	60	51
1941	388	53	47	10
1942	372	384	68	8
1943	432	58	107	16
1944	419	275	33	64
1945	268	210	91	7
1946	231	78	28	0
1947	338	313	55	53
1948	256	139	82	3
1949	249	211	48	4
1950	219	70	93	19
1951	248	34	51	0
1952	212	229	54	3
1953	145	515	40	2
1954	220	36	55	193
1955	181	126	302	218
1956	149	83	42	21
1957	180	192	82	395
1958	104	66	47	2
1959	183	58	25	24
1960	129	46	32	65
1961	149	51	52	46
1962	153	28	19	4
1963	165	31	39	11
1964	129	73	100	49
1965	149	296	119	75
1966	110	98	31	54
1967	88	114	34	18
1968	129	131	31	9
1969	131	66	297	256
1970	122	72	135	11
1971	122	156	74	8
1972	94	27	554	121
1973	124	87	148	5
1974	104	361	89	1
1975	92	60	113	53
Total (36 years)	7124	4892	3277	1879
Annual average	198	136	91	52

*Lightning totals through 1973 are from the National Center for Health
Statistics, Public Health Service; all other totals are from the National
Climatic Center, NOAA. Some overlap occurs in death totals from hurricanes and floods, because most hurricane deaths are from drownings
and because those occurring in rivers and streams swollen by hurricane
rains are also counted as flood deaths. Flood deaths include both river
and flash flood events. Lightning totals prior to 1953 include deaths
due to secondary effects of lightning, such as fires. (From H.M. Mogil
and H.S. Groper, Bul. American Met. Soc., Vol. 58, 1977.)

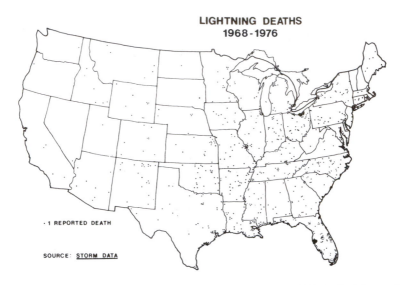

Figure 7-14. Lightning deaths are distributed throughout most of the United States. They are more common in the eastern half, however, as would be expected from the presence of more people and a considerable number of thunderstorms. (After Mogil, Rush and Kutka, *Weatherwise*, 1977.)

and hilltops (or the highest ground in the area), under isolated trees, near wire fences, and on tractors with implements in the ground.

Safer locations in the open are in valleys or under a small tree that is located a few hundred meters away from larger trees. Automobiles are ordinarily safe during lightning storms since they surround a person with metal and if they are struck the electricity flows through the metal to the ground.

A lightning strike does not always kill a person. One farmer who was plowing during a thunderstorm was struck and knocked unconscious. He was taken to a hospital where he survived, even though the lightning had burned a hole in his hat and had also burned a streak down his back. Recovery was similar to that following a stroke.

Lightning sometimes gives a warning a few seconds before it strikes. If your hair stands on end during a thunderstorm immediately squat as low as possible since this is an indication that you are about to be struck by lightning. If you lie flat on the ground the difference in potential between your head and feet may be enough to kill you if the strike is close to you.

It is possible to revive many people who have been struck by lightning and are apparently dead. A lightning strike is likely to stop the

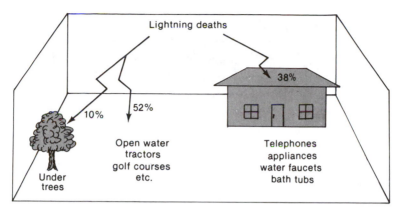

Figure 7-15. Unsafe locations during thunderstorms are under tall trees, in the open on tractors and golf courses, or in houses near appliances or plumbing.

heart. If there is no other damage to vital organs the victim may be revived by immediate application of cardiopulmonary resuscitation. This should be continued until an ambulance arrives.

Some people have survived more than one lightning strike. The Guinness Book of World Records lists a man who has survived five lightning strikes. During his job as a park ranger in Virginia he lost his big toe nail in 1942 to lightning, his eye brows in 1969, had his shoulder seared in 1970, and his hair set on fire in 1972 and 1973.

Lightning strikes have been known to be beneficial. In 1782 a paralyzed member of the household of the Duke of Kent was struck by lightning. He was immediately cured of the paralysis.

More recently in 1980 lightning struck Edwin Robinson of Falmouth, Maine who was blind and deaf from a head injury suffered in 1971. His sight and hearing returned slowly within a few months. When in New York for an appearance on ABC-TV's "Good Morning America" he said his scalp "felt funny, like whiskers on my face." The lightning strike was also changing his baldness to a thick head of hair at age 62.

SUMMARY

Benjamin Franklin greatly advanced our knowledge of electricity. He investigated thunderstorm electricity with kite experiments and determined that most thunderstorms have large negative charges at their base. He also developed the lightning rod that is still used for the protection of buildings.

Air discharges occur as the leader stroke beneath the cloud base fails to reach the ground. A bolt from the blue occurs if the air discharge travels for several kilometers and then reaches the ground farther away from the thunderstorm. St. Elmo's Fire occurs as the tops of tall objects become sufficiently charged by a thunderstorm overhead to emit a continuous glow. Ball lightning is another form of lightning that develops as a ball falls from the cloud base and explodes as it strikes the ground, or forms near a lightning discharge and floats through the air.

Thunder is produced by lightning discharge as the conducting channel of air is heated to five times the temperature of the sun in a very short period of time. The sudden expansion of the air in the conducting channel creates shock waves that are heard as thunder.

The lightning flash consists of several strokes. The first of these is a stepped leader that creates the conducting channel from the cloud base to the ground. After the conducting channel is developed a return streamer occurs from the ground to the cloud. The hot air channel conducts electricity because the molecules are ionized and therefore charged. A side effect of the extremely high temperature is the creation of a small amount of nitrogen fertilizer.

The development of electricity in clouds is associated with the raindrop formation process along with updrafts and downdrafts within the thunderstorm. Spontaneous freezing and shattering of supercooled water droplets helps initiate the charge separation within a thunderstorm. As the top of the storm becomes positively charged and the base negatively charged an induced charge occurs on all of the ice particles and water droplets within the central part of the thunderstorm. The small cloud droplets within an intense updraft bleed off the negative charges from the lower side of larger ice pellets. The concentration in the upper part of the thunderstorm of smaller cloud droplets that are positively charged intensifies the charge separation process.

Lightning is responsible for almost 200 deaths per year in the United States and about 25 million dollars worth of damage per year. Properly installed lightning rods are effective in protecting buildings from lightning damage. Over one-half the lightning deaths occur outdoors with dangerous locations being near open water, on a tractor with an implement in the ground, or on a golf course with a golf club over your head. Other unsafe locations are under large trees and indoors near telephones or metal appliances.

8
Ice from the Sky

At first I thought it was only a bad dream, but as I awoke more fully I realized that the pounding sounds on the roof were real. The continuous noises were punctuated by intermittent loud thumps. Occasional lightning flashes revealed the accumulating hailstones on the lawn. Most were golf ball size but scattered among them were chunks of ice as large as baseballs. One of them appeared to be larger than a softball!

The sound of breaking glass drew me to the kitchen where a large chunk of ice lay on the floor among pieces of the broken west window. The hailstone was larger than my closed fist and contained knobs all over it similar to my knuckles. After placing the hailstone on the hearth one blow from a hammer shattered it into several pieces that showed an interesting pattern of alternating layers of ice within it. My interest in this was dampened, however, since I did not need a vivid imagination to know what such hailstones were doing to my new car parked outside. . . .

LOSSES FROM HAIL

In some parts of the United States hailstorms frequently damage personal property, such as automobiles and houses, in addition to the effect of large hailstones on plant and animal life. Two persons have been killed in the United States by large hail. A farmer near Lubbock, Texas, was caught outside on May 13, 1930, during a hailstorm and was beaten so severely by large hail that he died within a few hours. Softball size hail killed a baby in Fort Collins, Colorado, July 30, 1979. This storm injured 25 others and extensively damaged 2500 automobiles and 2000 houses. This hailstorm alone brought damages that totaled $20 million. Reports indicate that many people have been killed in India by hailstorms. Hailstones are also occasionally large enough to kill cows and horses; sixteen horses were killed in South Dakota on July 5, 1891, by very large hailstones.

Hail damage to agricultural crops is extensive in the United States (Figure 8-1) and amounts to about $680 million annually. This represents about 2% of the nation's crop production. In some areas of the Great Plains about 20% of the value of crops is lost to hail annually. Hail insurance on the 15% of the nation's crop covered by hail insurance costs about $300 million annually.

Figure 8-1. Hail can completely destroy agricultural crops. A corn crop such as that shown above would be unable to recover from the effects of a hailstorm.

Personal property damage is not as extensive as crop damage but amounts to about $70 million annually (Figure 8-2).   Large cities occasionally suffer extensive damage from hailstorms. Oklahoma City received $20 million worth of damage from hailstorms on May 23 and 24, 1968.   Damage from large hail in Amarillo, Texas, on June 17, 1969, totaled $15 million with 25,000 homes and 6,000 automobiles severely damaged by the storm.

OCCURRENCE OF HAILSTORMS

The distribution of hailstorms across the United States is slightly different from the distribution of thunderstorms and tornadoes. The area of maximum hail might be expected to be the same as the area

Figure 8-2. Large hail is also a threat to various types of personal property including houses, automobiles, and aircraft. The severely battered wings of this airplane show the effects of flying through a thunderstorm that contained many hailstones. (Courtesy of NOAA.)

of maximum thunderstorms, but this is not the case. The maximum occurrence of thunderstorms is in Florida, with a secondary maximum in the central United States in southern Colorado and New Mexico, while the maximum occurrence of hail is in northeastern Colorado and southeastern Wyoming (Figure 8-3). As many as 8 days per year on the average have hailstorms in this location, while hail occurs an average of four days per year in the area encompassing parts of South Dakota, Nebraska, Kansas, Oklahoma, Texas, and New Mexico. Hail very rarely occurs in the Gulf Coast and Florida area, in comparison to the large number of thunderstorms there. This shows that specific conditions must exist in a thunderstorm to produce hail.

The season of major hailstorm occurrence is related to the location of the jet stream. Hailstorms start to become a problem in Texas and Oklahoma during April. By mid-May they are a greater threat northward over Kansas and Colorado, and during June hailstorms are more likely in Wyoming, South Dakota, and Montana. The region of major hailstorm activity corresponds to the migration of the jet stream. Hailstorms are more likely in the spring farther south with the band

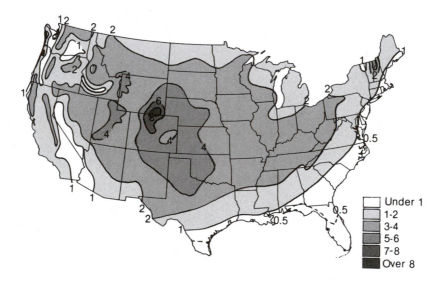

Figure 8-3. Hailstorms are most common in an area extending from Texas through South Dakota. They occur most often in Northern Colorado and Southeastern Wyoming. Because of the intensity of agricultural activities western Kansas suffers more damage to agricultural crops than other locations. (After Environmental Data and Information Services, NOAA.)

of greatest occurrence moving to the northern United States during the summer as the jet stream follows this pattern. Hailstorms are similar to tornadoes in this respect, since they also follow the movement of the jet stream.

Hailstorms over the whole United States are more abundant in May and June than other months of the year (Figure 8-4). They also occur

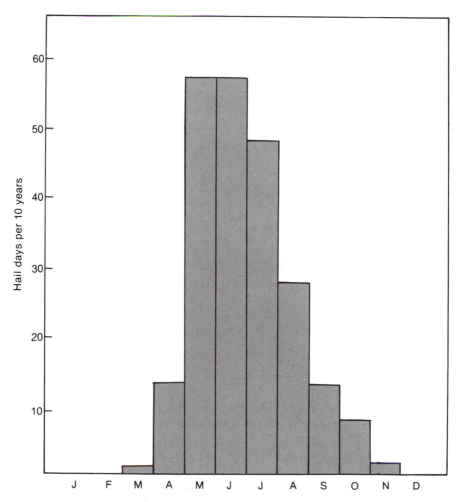

Figure 8-4. The monthly distribution of hailstorms shows they are more common in May and June with considerable activity also in July and August. Severe thunderstorms are very uncommon during the winter months.

frequently in July and August making late spring and summer the time of major hailstorm activity. Since hailstones are composed of ice they might be expected to be more common in the wintertime when it is colder, but this is not the case. Large cumulonimbus with high velocity updrafts are required to form large hailstones and these occur in the warm rather than the cold season. Thunderstorms develop more frequently as surface heating causes the air to become buoyant and rise to great heights resulting in hail formation. The most frequent time of hail is 3 to 4 pm with almost all hailstorms occurring between 2 and 6 pm. This emphasizes the importance of surface heating and the development of large thunderstorms in the atmosphere.

Hailstones have accumulated on the ground to depths of more than 30 cm. Hail fell at Seldon, Kansas, from 5:15 to 6:40 pm on June 3, 1959, with accumulations of marble-size hail to a depth of 45 cm. Thunderstorms that produce heavy hail also frequently produce considerable rain. Runoff from the storm may accumulate hailstones into drifts that have grown to depths of more than 2 m.

CHARACTERISTICS OF HAILSTONES

The appearance of a hailstone that is cut in half is similar to an onion, since it is composed of several layers (Figure 8-5). The layers of hailstone consist of alternating milky opaque ice and clear ice. The environment within the thunderstorm determines the appearance of the ice within a hailstone. If the water freezes very slowly, bubbles have time to get out of the water producing a layer of clear ice. Rapid freezing traps the bubbles within the ice producing an opaque milky layer. Rapid freezing occurs at cold temperatures high in the thunderstorm. Snowflakes and ice crystals are also likely to be present at this level. Some of these become entrapped within the outer layer as it forms on the hailstone contributing to the milky, opaque appearance of the ice formed high above the freezing level. Thus, the alternating rings in the hailstone correspond to the regions in the thunderstorm where the ice was formed above and below the freezing level.

Small hailstones are usually round in shape, but large hailstones are seldom round and smooth. Large hailstones are often knobby and elongated; oblate spheroids are a common shape. Sometimes large stones are conical shaped or flat. The most typical shape of a large hailstone is an oblate spheroid that has a very uneven and knobby outer layer.

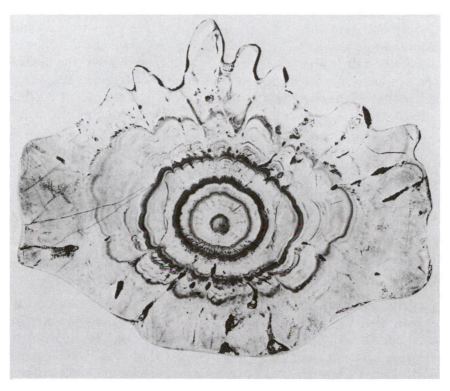

Figure 8-5. This photograph of the largest hailstone ever recorded shows five alternating layers made up of clear or milky appearing ice. This hailstone fell near Coffeyville, Kansas and measured over 19 cm (7 1/2 in) in diameter. (Courtesy of NCAR.)

Studies conducted in wind tunnels with ice suspended in a stream of high speed air show some of the reasons for the different shapes of hailstones. The dynamics of airflow around chunks of ice help determine the various shapes. The optimum shape of a hailstone is related to its size and orientation within the updrafts with the dynamics of airflow dictating that a flat shape is more likely at one size and a conical shape at another. The rough outer layer of large hailstones develops as masses of water and ice accumulate and freeze because of the lower temperature of the ice as it falls from the very cold atmosphere above.

The size of hailstones ranges from 0.5 cm to 19 cm. The largest hailstone that has been documented fell near Coffeyville, Kansas, on March 9, 1970, and measured 44 cm in circumference, 19 cm in diameter and weighed 766 g (Figure 8-5). The size distribution of hail is slightly different in various geographical locations. Some studies (Figure 8-6) have shown that the average hail size is one to two cm in Denver while smaller hail is more common in New England. The size distribution of hail in Canada is not too different from that in Denver, but hailstones are slightly smaller. The most frequent size of hail is pea-size. Hailstones the size of grapes fall quite frequently, while walnut size hail falls less frequently. Still less often golf ball, tennis ball, baseball, and softball size hail is generated by thunderstorms.

The eastern United States occasionally has large hail, but more often hailstones are small. This is related to the size and severity of the thunderstorm. Great Plains thunderstorms are generally larger and capable of producing the largest hailstones. The largest recorded hailstone, prior to 1970, fell near Potter, Nebraska, on July 6, 1928, and was 14 cm in diameter.

UPDRAFT SUPPORT OF HAILSTONES

Large thunderstorm updrafts are necessary for the growth of hailstones. If the upper atmospheric temperature is cold, surface heating of the lower atmosphere causes the atmospheric stability to decrease creating an environment for the development of large thunderstorms with intense updrafts. The updrafts necessary to support hailstones of various sizes have been calculated. Tremendous updrafts are necessary to support the largest hailstones. Information on the mass per unit volume (density) of hailstones is required in order to calculate the magnitude of the supporting updraft. The density of normal ice is 0.9 gm/cm^3, but hailstones have varying densities depending on the number of milky layers and number of bubbles that are incorporated into them. The average density of hailstones is about 0.8 gm/cm^3.

An updraft of 95 km/h would be required to support a hailstone with a diameter of 2.5 cm. A hailstone of 8 cm in diameter requires an updraft of 200 km/h to support it (Figure 8-7) while a 13 cm hailstone would require an updraft velocity of 377 km/h for support. A 19 cm hailstone would not have to be completely supported by an updraft since the last ring could grow as it fell from the cold air above the freezing level to the surface. However, the hailstone would have to be supported by an updraft of sufficient magnitude to carry it upward when its size corresponded to the next to last layer with-

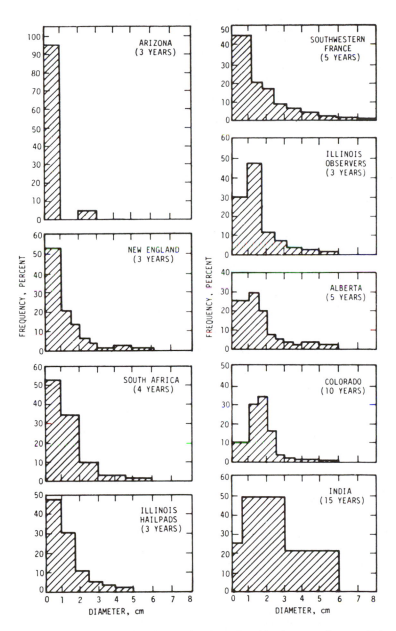

Figure 8-6. The distribution of hailstone sizes in various locations reveals some slight variations. Typical hailstone sizes vary from 1 to 2 cm in diameter. (After Changnon, *Journal of Applied Meteorology,* January 1971.)

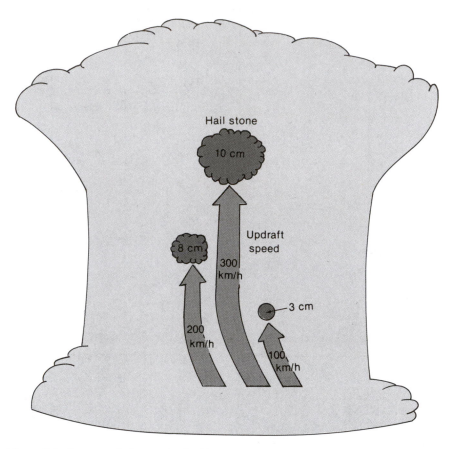

Figure 8-7. Strong updrafts are required in thunderstorms to support large hailstones. An updraft of 300 km/h (180 mi/h) would be required to support a hailstone 10 cm (4 in) in diameter.

in the hailstone. This layer of larger hailstones indicates a considerable mass of ice that would require very large updrafts within the thunderstorm.

ATMOSPHERIC CONDITIONS DURING HAILSTORMS

The atmospheric conditions necessary to generate hailstorms are the subject of scientific investigations since all the factors that cause hailstorm formation have not been determined. One of the important contributing factors is the presence of an incoming cold front. Major hailstorms usually occur in the warm air preceding the movement of a mass of colder air into a locality (Figure 8-8). Surface streamline confluence near a cold front contributes to atmospheric instability. The jet stream and upper atmospheric streamline diffluence are also associated with hailstone formation. The jet stream is influential in providing flow of air around a large thunderstorm feeding energy into the storm that becomes organized into high speed updrafts and downdrafts. Upper atmospheric streamline diffluence is important in contributing to the updrafts within a thunderstorm. Therefore both surface and upper atmospheric features are related to hailstorm formation.

Considerable wind shear is normally present in the atmospheric environment where hailstorms form. A study of Modahl in 1969 showed that the wind speed from the surface to 500 mb was very different from that compared to the wind speed from 500 mb to 250 mb. The difference between the two was related to the size of the hailstones produced. A wind shear of 65 km/h between the two layers corresponded to heavy hail, 61 km/h to moderate hail, and 48 km/h to light hail. A wind shear of 44 km/h or less was related to no

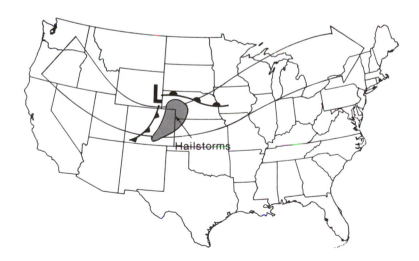

Figure 8-8. Hailstorms are most common within the warm air sector of a frontal cyclone. They are also usually associated with strong winds in upper levels of the atmosphere.

hail formation. This showed a direct relationship, with greater wind shear more likely to produce large hail.

An unstable atmosphere is associated with hailstorms. A comparison of the atmospheric temperature profile near the time of different hailstorms, as well as the temperature profile near the time of tornadoes, has shown that hailstorms are associated with an unstable atmosphere, although an upper air inversion is not usually present when hailstorms develop. This indicates one of the atmospheric conditions that causes hailstorms to have a different geographical distribution from that of tornadoes.

HAIL PRODUCING THUNDERSTORMS

Hailstorms may be associated with isolated supercell storms, multicell thunderstorms, or squall lines. There is some experimental evidence that indicates that the most severe cell in a squall line or in a multicell group may develop the same organized structure as the single supercell thunderstorm. Most thunderstorms that produce tornadoes also develop hail, although many hail producers do not generate a tornado. The appearance of a thunderstorm that has hail is different from an ordinary thunderstorm. The optical characteristics of sunlight striking a hailstorm cause a dark blue-green color. The very large water droplets and hail in the thunderstorm reflect the light in a different way causing the greenish appearance.

An analysis of hailstorms in South Dakota by Dennis showed that the average top of the thunderstorm was about 14 km for 134 hailstorms. If a thunderstorm grew to a height of 13.7 km, the probability was 50% that it would produce hail. If the thunderstorm grew to 17 km in the late afternoon, it was almost certain to have hail associated with it. Thunderstorms developing to heights above 14 km near noon were almost certain to have hail associated with them.

The height of thunderstorms that produce hail varies with geographic location (Figure 8-9). A thunderstorm that reaches only 8 km in Alberta, Canada, has a 50% chance of producing hail while a height of 17 km in Texas results in a similar probability. Hailstorms originating in locations with warmer and more moist subcloud air must reach greater heights to produce hail at the ground.

A severe thunderstorm produces rain, hail, and tornadoes in specific areas within the thunderstorm. Hail normally falls in the cen-

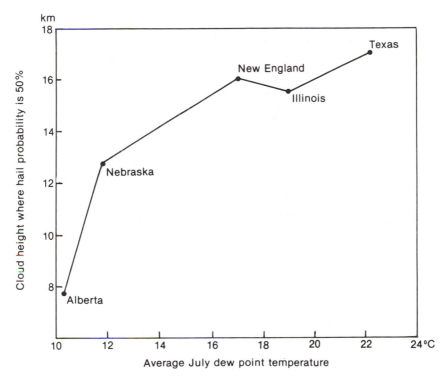

Figure 8-9. Hail is more commonly associated with large thunderstorms. Those that produce hail are typically less than 8 km in height in Alberta while they are typically more than 17 km in height in Texas. (After Grosh and Morgan, Preprint Volume, Severe Local Storms, AMS, 1975.)

tral part of a thunderstorm with the major rain area in the leading part of the thunderstorm, typically northeastern, and tornadoes in the southern part of the thunderstorm. The major hail area in the central part of the storm is related to the thermal updraft and the flow of air within the thunderstorm, causing a hailstone to take one or more paths above and below the freezing level (Figure 8-10).

As in the case of tornadoes, most thunderstorms that produce hail move toward the northeast. Occasionally, hailstorms move toward the south or southeast. Most hailstorms come from the southwest for the same reason that other severe thunderstorms move in this direction. The upper atmospheric winds are more likely to come from this direction, thereby, determining the path of the thunderstorm. The speed of travel of hailstorms is, in general, slower than

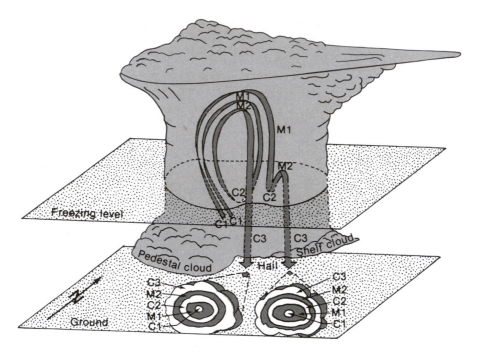

Figure 8-10. The various rings of a hailstone are formed as it circulates within a thunderstorm. Clear layers are formed where the ice freezes very slowly; milky layers are formed under more rapid freezing conditions. The number of trips through a thunderstorm determines how many rings of ice the hailstone will have. (After Eagleman, *Meteorology: The Atmosphere in Action*, D. Van Nostrand Company, 1980.)

the average winds. Slower moving thunderstorms are likely to have hail associated with them if they are traveling toward the northeast.

HAIL FORMATION

The nature of hailstone generation by thunderstorms is still under investigation. Hail may be generated by a combination of updrafts and downdrafts in a double vortex thunderstorm. Hail is generated between the two vortices by single or repeated circular trips within the thermal updraft of the thunderstorm that carry the hailstone above and below the freezing level (Figure 8-10). Time spent above the freezing level results in a milky hail layer from rapid freezing, while the hailstone develops a layer of clear ice below the freezing level. Additional layers may be deposited as the thermal updraft carries large quantities of supercooled liquid water above the freezing level to coat the large hailstone. The number of trips above and below the freezing level or number of coatings of supercooled water determines the number of layers within the hailstone giving general information on the path of the hailstone within the thunderstorm.

A large hailstone that makes only one trajectory through the thunderstorm can develop successive layers near the freezing level if turbulent motion carried the hailstone repeatedly across the freezing level producing successive layers of milky and clear ice. This possibility combined with coatings of supercooled liquid water delivered above the freezing level by the thermal updraft also explain the layered nature of a hailstone.

Frequently the outer ring of hail is much larger than the other inner layers. This shows that even fairly large hail can stay in the upper part of the thunderstorm for some time acquiring alternating rings of ice. The last clear layer with the greatest mass of ice develops as the hailstone falls through the part of the cloud with high water content. A large hailstone may be cold enough to cause freezing of liquid water for some distance below the freezing level.

HAIL RESEARCH

The Illinois State Water Survey has investigated hail occurrence for many years. A network has been established over several counties with various instruments providing information. Considerable information is collected by hailstools that are dented when struck by hailstones (Figure 8-11). The shape of the hailstool and location of the indentations can be used to calculate the fall angle of hailstones. A dent in the base of the stool can only be produced by a hailstone coming from a direction that allows it to miss the top ring. The angle of fall and the size and degree of indention can be used in determining the energy of hail. The size, angle of fall, and the impact of hailstones provide information on the damage to be expected from hailstones of various sizes.

A dense network of recording rain gauges combined with hail instruments has been used to measure various characteristics of Illinois hailstorms. A hailstorm in 1968 that crossed the network was generated by a typical synoptic situation (Figure 8-12). A low pressure center traveled from Nebraska through Iowa, and occluded in Illinois, generating large thunderstorms, hail, and tornadoes. The 500 mb map on the 15th of May, 1968, had very strong streamline diffluence. This was a contributing factor to the development of the large thunderstorms.

Figure 8-11. Instruments that measure the characteristics of hailstones may consist of a hail stool (left hand photograph) or more elaborate instruments that record the impact of hailstones and other information. (Courtesy of Stanley Changnon, Illinois State Water Survey.)

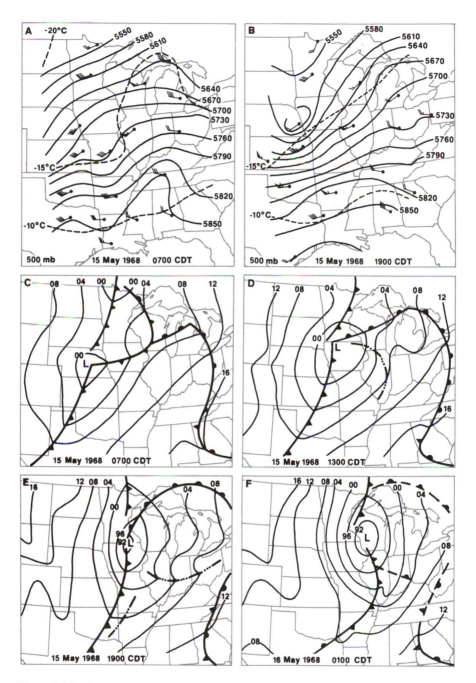

Figure 8-12. The synoptic maps above correspond to a major hail outbreak in Illinois. The surface maps show a frontal cyclone moving through Nebraska, Iowa, and Wisconsin. Severe thunderstorms developed within the warm air sector beneath an area of upper air divergence as shown by the airstreams on the 500 mb maps. (After Changnon and Wilson, Report 68, Illinois State Water Survey, 1971.)

Several storms crossed the measurement network from this weather system. A steady state thunderstorm left 130 mm of rainfall along one strip. A swath 30 km wide and 80 km long received heavy hail damage. Five tornado damage paths were produced by the thunderstorms and several smaller hailswaths.

Information on the duration of hail fall showed that in some locations it lasted for 40 min, while in others it varied down to 10 min. The largest hailstones from this storm were 5 cm in diameter in the central part of the network, while very small hail occurred in surrounding areas. The relationship between the location of tornadoes, rainfall, and hail swaths is shown in Figure 8-13. Crop damage was almost complete within the central part of the hail swath with varying degrees of damage outward.

There is considerable interest in hail suppression because of the extensive damage caused by hailstorms each year. Although many conceivable methods of modifying hailstorms could be used, the only extensively used technique is cloud seeding with silver iodide to alter the ice nucleation process within the thunderstorm. Most operational cloud seeding programs are based on the concept of competitive embryos, while a few are based on the glaciation concept (Figure 8-14). Glaciation cloud seeding involves heavy seeding of silver iodide within the cloud in an attempt to convert all the supercooled cloud droplets into ice crystals, thereby eliminating the opportunity for hailstone growth. This seeding method is generally considered to be impractical because of the large quantities of silver iodide required for glaciation of a large thunderstorm.

The beneficial competition cloud seeding approach is based on the concept of introducing enough silver iodide into the cloud to create a large number of ice crystals and hailstone embryos resulting in such great competition for the liquid water supply that none are able to grow into large damaging hailstones.

The Russians have reported very good results in hail suppression using rockets that deliver silver iodide into the hailstorm. Hail suppression in the United States has not been so encouraging, although operational hail suppression is used over an area of about 170,000 km^2 in the Great Plains.

Some positive and some negative results have been obtained from hail suppression experiments. The 4 year statewide hail suppression program in South Dakota was ended in 1979 by political contro-

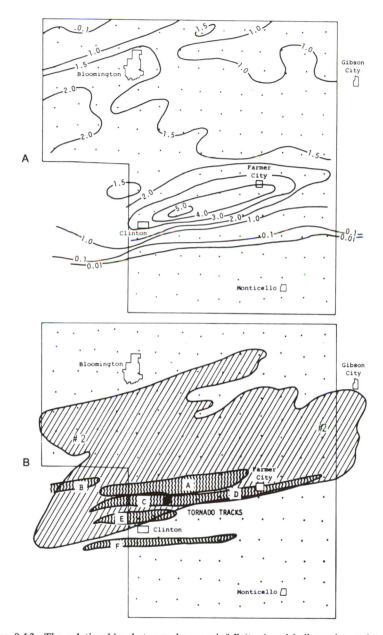

Figure 8-13. The relationships between heavy rainfall (top) and hail swathes and tornado tracks are shown in this illustration. Tornados generally occur in the southern part of a thunderstorm with most intense hail near the central part and heaviest rain in the leading part of the thunderstorm. The hail swath shown by the area containing slanted lines was more than 15 km in width and more than 30 km in length. (After Changnon and Wilson, Report No. 68, Illinois State Water Survey, 1971.)

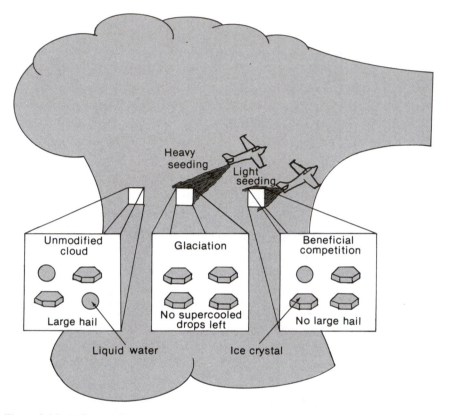

Figure 8-14. Hail suppression experiments may involve the glaciation concept where heavy seeding is used to convert all of the supercooled liquid water droplets to ice crystals. This may prevent hailstone formation. The other concept involves beneficial competition in which a certain percentage of the supercooled liquid water droplets are converted to ice crystals so that the number of hail embryos is increased to such a level that no extremely large hail are formed while many smaller hail may result.

versy. The National Hail Research Experiment (NHRE) conducted for 3 years (1972-74) in northeast Colorado by the National Center for Atmospheric Research with funds from the National Science Foundation has been interrupted, if not terminated. This research project was aimed at gaining enough information on the techniques and results of hail suppression to allow operational hailstorm modification programs to proceed with greater certainty of the expected modifications.

In spite of the uncertain technology many user groups, especially those connected with agricultural production, desire to continue operational hail suppression programs. Some of these seem to be effective. An evaluation of five programs that have been in operation for at least 3 years in one locale (Table 8-1) indicates reductions in crop loss from 48% in western Texas to 20% in South Dakota. These used the beneficial competition approach with aircraft seeding the main updraft near the cloud base except for the South Africa program, which used the glaciation concept.

It should also be pointed out that a few experimental programs indicate an increase in hail from cloud seeding. One analysis of the NHRE data indicated an increase in the total number of hailstones.

TABLE 8-1. Current best estimates of hail suppression effects in operational programs. (After Changnon, S.A., "On the Status of Hail Suppression," *Bulletin Amer. Met. Soc.* Vol. 58, 1977).

SEASONS (NO)	AREA	SEEDING METHOD	EFFECT (%)	BASIS
		Western Texas		
4	Two counties	Cloud base	−48	Crop loss cost[a] compared to prior history.
		Southwestern North Dakota		
15	Two counties	Cloud base	−31	Crop hail insurance rates compared to adjacent unseeded counties.
7	Four counties			
		North Dakota pilot project		
4	One county	Cloud base	−30	Composite of hail characteristics, radar, and crop loss data.
1	Three counties			
		South Africa		
4 ½	5300 hectares	Cloud top	−23	Crop severity ratio[b] compared to previous history.
		South Dakota		
4	50 to 70 percent of state	Cloud base	−20	Crop loss cost compared to prior history.

[a]Crop loss cost = (dollar loss/dollar liability) x 100%.

[b]Crop severity ratio is the average percent loss over the area hit.

This project has also indicated that the results of hail suppression programs may depend on the time of seeding and choice of clouds that are seeded. There are indications that hail production is increased if seeding is conducted after a thunderstorm has reached the supercell stage. It may be more effective to seed developing thunderstorms as they appear to be more responsive. The supercell thunderstorm, after achieving a steady state internal structure has an abundant supply of supercooled water in the region of the strong updraft (Figure 8-15). The hail embryos that become the largest hailstones are carried over

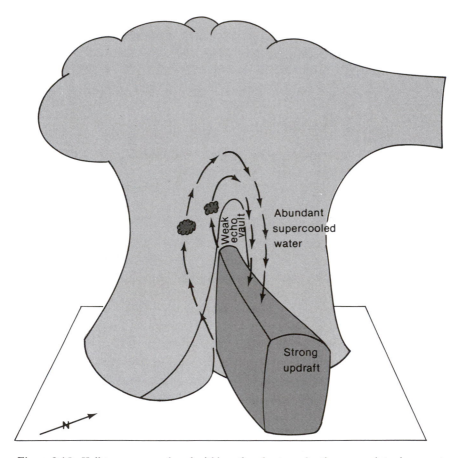

Figure 8-15. Hailstones are produced within a thunderstorm by the appropriate air currents. Within a double vortex thunderstorm the primary path of hailstones is probably between the vortices within the strong updraft in the leading front part of a thunderstorm. As a hailstone falls into the rain area, abundant supercooled water is available for making it larger.

the top of the weak echo vault near its boundary where they are able to take advantage of the abundant supply of supercooled liquid water in the main updraft. Seeding of the updraft of a supercell thunderstorm may generate more hail because of the almost unlimited supply of supercooled liquid water after the thunderstorm has become organized into a steady state circulation.

The more common multicell hailstorm may respond differently to cloud seeding. However, the NHRE, which included a randomized cloud seeding experiment to gather information on the effects of seeding various types of clouds, was suspended after three years of data gathering with results that are inconclusive. They did not indicate with any certainty whether the cloud seeding reduced hail, had no appreciable effect, or resulted in more hail than would have fallen naturally.

SUMMARY

Hailstorms cause considerable damage to agricultural crops each year in addition to damage to personal property and animal life. Hailstorms are the greatest problem in the central Great Plains reaching a maximum in Northeastern Colorado and Southeastern Wyoming. Hailstorms are most frequent in May and June in the late afternoon hours.

Hailstones are composed of alternating layers of milky opaque ice and layers of clear ice. The thunderstorm environment where the hailstones originated determines the appearance of the ice layer. Hailstones come in various shapes. High speed updrafts are required in thunderstorms to form large hailstones. An updraft of 100 km/h is required to support a hailstone of only a few centimeters in diameter.

Some important atmospheric conditions that contribute to the development of hailstones are surface streamline confluence ahead of a cold front, high wind speeds within the jet stream, large wind shear and an unstable atmosphere. Hailstorms may occur from any of the three different types of thunderstorms—supercell, multicell, or squall line. Multicell and squall line thunderstorms account for a greater percentage of the total amount of hail, but the hail from a supercell thunderstorm may be larger and more damaging. Hail is most likely to fall beneath the central part of a thunderstorm while heavy rain generally occurs beneath the leading edge of the thunderstorm, and tornadoes, if present, are more likely on the southern edge of a thunderstorm moving toward the northeast. Most hailstorms come from the southwest and travel at a speed slightly slower than the average wind throughout the atmosphere.

Various types of hail research are in progress. Hail stools and recording rain gauges are used to give information on the energy and distribution of hail and rain within a thunderstorm. Considerable interest in hail suppression has existed since the Russians described tremendous success in reducing crop loss from hail during the 1960s. Several operational hail suppression programs in the United States indicate good results with reductions in crop loss of as much as 48%. Other experiments have given smaller reductions in crop loss and in some cases even indicate an increase in the amount of hail.

Most of the cloud seeding programs are based on the concept of competitive embryos in the hail development process. The other concept used in operational cloud seeding programs is that of glaciation. This involves heavier silver iodide seeding rates in an effort to convert all the supercooled cloud droplets into ice crystals.

The National Hail Research Experiment conducted for several years by the National Center for Atmospheric Research has failed to provide convincing evidence that cloud seeding is effective in reducing the amount of hail. Interest in hail suppression is likely to continue, however, with attempts to refine the seeding techniques and selection of clouds to be seeded, since a tremendous potential exists for preventing the damage that results from hailstorms.

Part 2
Hurricanes and
Other Unusual Weather

9

The Mighty Middle-Size Storm

A cooperative weather observer, J.E. Duane, located on one of the Florida Keys experienced the hurricane on September 2, 1935, that contained extremely strong winds and the lowest pressure ever measured in the United States. He gave this account:

The barometer reading was decreasing on September 2, as large sea swells and heavy rain continued throughout the afternoon. The temperature was 79°F when the winds increased to hurricane speeds at 6:45 pm. As we crouched in our cottage, a huge 15 by 20 cm beam 6 m long came flying through and barely missed three persons. At about 9 pm the winds abated and we could hear other noises besides the wind and rain. The barometer stood at a low reading of 27.2 in and the light winds told us that we were in the eye of the hurricane. We headed for the last and only cottage that I thought could stand the blow due to arrive shortly. All 20 of us now waited patiently for the hurricane winds to return.

During the lull of 55 min, the skies cleared, with a very light breeze throughout the lull. As I took my flashlight to investigate the sea level only about 20 m from the cottage, I saw the water begin to rise very rapidly. As I raced back to regain entrance to the cottage, the water caught me waist deep. I managed to get back into the house and close the door. But soon the house began to sway back and forth and we knew it was floating. At 10:15 pm a blast of wind came at full force from the south-southwest. The house began to break apart. I glanced at the barometer which read 26.9 in, but dropped it into the water as I was blown out to sea. I became entangled in the fronds of a coconut tree and hung on for dear life. I could see the house floating out into the ocean as those still inside flashed their flashlights. I was then struck by some object and knocked unconscious.

I became conscious in the tree a few hours later at 2:25 am on September 3. I was 7 m above the ground and all the water had disappeared from the island. I climbed down from the tree and saw that the house had been blown back onto the island. I was amazed to find all the people safe inside. A barometer in the house was showing signs of rising pressure, but very slowly. Hurricane winds continued until 5 am with terrific lightning flashes. Although the whole camp was demolished not a single life was lost. . . .

HURRICANES, TYPHOONS AND CYCLONES

Hurricanes are strong atmospheric vortices that are intermediate in size between the larger frontal cyclones and much smaller tornadoes. These tropical storms are called cyclones in the Indian Ocean and typhoons in the Pacific. They originate only in the tropical trade-winds where the ocean temperatures are quite warm. The tropical tradewinds are affected by large high-pressure areas, subtropical highs, that are present most of the time over the oceans at about 30° latitude. In the northern hemisphere these are called the Hawaiian and Bermuda highs. The winds blowing around the high pressure travel anticyclonically in the northern hemisphere and create persistent northeast winds on the southern side of the high pressure areas that are known as the tradewinds. It is within these tradewinds where a cloud mass may be transformed into one of the most devastating storms on earth, not as strong as a tornado but much larger. Hurricane watch areas cannot be forecast on the basis of appropriate atmospheric conditions, as tornado watches can, since the precise atmospheric conditions that cause the development of hurricanes are not known.

STAGES OF HURRICANES

The beginning of a hurricane is a small tropical disturbance in the tradewinds. Such tropical disturbances are often present because the high moisture content of the air and the abundance of heating from the ocean surface cause the air to become moist and unstable enough to rise and form clouds. The normal straight flow in the tradewinds may become curved and form so-called easterly waves with areas of cloud development. As these develop into vortices, the formation of a hurricane is in progress (Figure 9-1). It is not known why some tropical disturbances suddenly grow into a tropical storm and others do not, even though preliminary conditions appear similar.

As the developing hurricane passes to the next stage, called a tropical depression, the central pressure begins to fall and the winds begin to flow circularly around the lower pressure (Figure 9-2). An eye begins to appear characterized by little or no clouds, warmer temperatures, and lighter winds. One or more closed isobars are typical of this stage with winds less than 65 km/h.

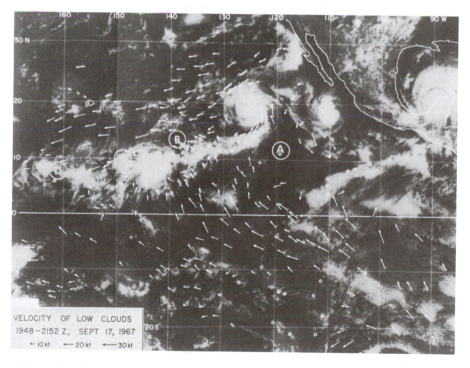

Figure 9-1. The birth place of hurricanes is within the tradewinds over tropical waters. Tropical disturbances and hurricanes in various stages of development are shown here on the satellite photograph (above) and analysis of wind direction (p. 166). (After ESSA, Technical Report No. 51, 1969.)

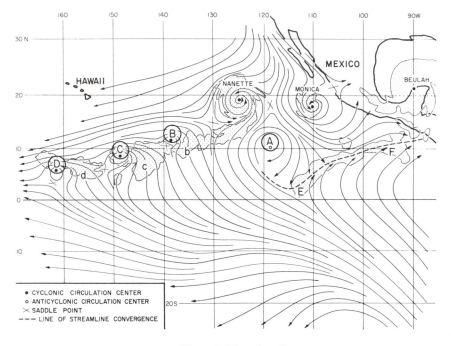

Figure 9-1 (continued)

In the third stage the developing hurricane becomes a tropical storm with more closed isobars and lower central pressure with winds gaining speeds of 65 to 120 km/h. The hurricane stage is reached as the winds become greater than 120 km/h. The central pressure of the hurricane is typically less than 950 mb as the storm matures to its full strength.

ORIGIN OF HURRICANES

Hurricanes are generated as the atmosphere absorbs large quantities of moisture from the ocean. An upper air inversion is usually present over large areas of the tradewinds as hurricanes develop. The inversion is associated with sinking air around the large subtropical high. This contributes to the buildup of a large contrast between the warm, moist air below the inversion and the cooler air aloft. Atmospheric vortices frequently develop near the intertropical convergence zone, a region where the tradewinds of the northern and southern hemispheres meet. A very interesting satellite photograph was taken

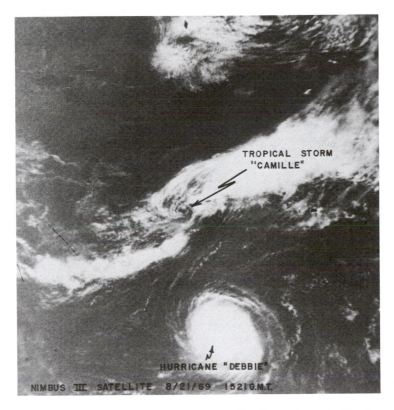

TROPICAL STORM
"CAMILLE"

HURRICANE "DEBBIE"

NIMBUS III SATELLITE 8/21/69 1521 G.M.T.

Figure 9-2. The developing stages of tropical storm Camille are shown in this photograph. Hurricane Debbie is located southward from the developing tropical storm.

on the 26th of September, 1976, that shows four vortices present at the same time (Figure 9-3). It is not unusual for two or three hurricanes to exist together in the same ocean. The formation of a hurricane, like other atmospheric vortices, involves the combined effects of pressure and circular winds (Figure 9-4). As the winds become circular around an area of low pressure, the inflow of air toward the low pressure center is prevented. The spiraling winds then form a vertical cylinder extending upward through the atmosphere. Thus, the pressure inside the cylinder at the surface is more like that above than that at the surface on the outside of the cylinder.

An atmospheric vortex can be driven by an outflow of air and associated low pressure at the top of the vortex, as in tornadoes and frontal cyclones, or by the spiraling winds themselves as in hurricanes.

Figure 9-3. Several vortices are developing within the tradewinds in this satellite photograph taken on September 26, 1976. Two of the storms reached hurricane intensity. The one on the right became Hurricane Liza which struck Lapaz and caused millions of dollars worth of damages.

As the cloud bands around the eye of the hurricane produce large quantities of condensed water, the latent heat of condensation (575 cal/g) keeps the rising air currents warm enough to rise even faster and add to this atmospheric heat engine process (Figure 9-5). The rest of the heat engine is supplied as the sun adds heat to the ocean water and evaporates some of it. The resulting warm moist air feeds the hurricane as it is pulled toward the low surface pressure. The importance of this part of the heat engine is emphasized by the fact that hurricanes originate only over warm oceans and decay rapidly as they move over land. Hurricanes do not form at the equator even

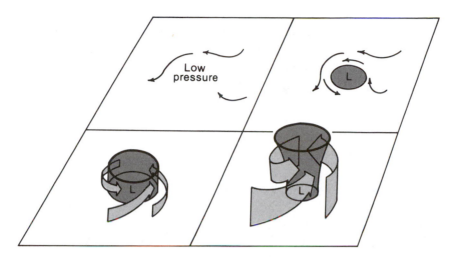

Figure 9-4. As the pressure begins to fall over tropical waters, the air begins to circulate around this area of lower pressure. As the wind speeds increase, the centrifugal force directed outward from the low pressure center increases making it more difficult for air to move into the low pressure center. As this pressure–wind speed relationship intensifies a hurricane is formed.

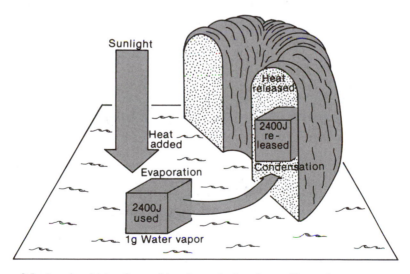

Figure 9-5. A major driving force of hurricanes is the release of latent heat as water vapor condenses within the cloud bands. This sets up a natural heat engine as sunlight evaporates water with the water vapor condensing within the wall of the hurricane. This causes the air to rise faster and is a major driving force for the hurricane.

though ocean temperatures are high and the atmosphere is very humid. The region of hurricane formation is from 5 to 30° north and south latitude (Figure 9-6). This region is not only characterized by warm ocean temperatures, but also has sufficient Coriolis acceleration to cause the storm to rotate. At the Equator, the only force due to the earth's rotation is a centrifugal force, whereas away from the equator toward the poles there is considerable spin about a vertical axis because of the earth's rotation. This spin is necessary to start the circular winds of the hurricane.

The formation region of most hurricanes that affect the United States is the Gulf of Mexico or the Atlantic Ocean. Occasionally, hurricanes form in the Pacific Ocean and hit the West Coast. The paths of typhoons, cyclones, and hurricanes are similar in the northern hemisphere (Figure 9-7). China and Japan are affected by many typhoons. The Indian coastal regions around the Bay of Bengal have been greatly affected by tropical storms because of the density of population on the coast.

In the Southern Hemisphere, tropical storms travel in an opposite direction to those in the Northern Hemisphere. The direction of the surface tradewinds combined with the winds above influence the direction of a hurricane. As they migrate northward, they are influ-

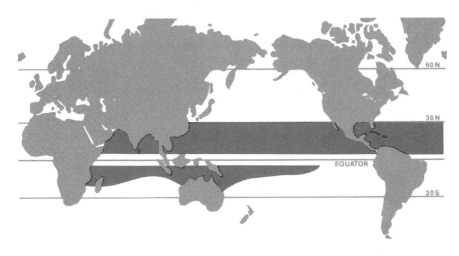

Figure 9-6. The hurricane formation region is governed by the location of warm ocean temperatures greater than 26°C. In addition hurricanes do not form within about 5° of the Equator because of lack of Coriolis acceleration.

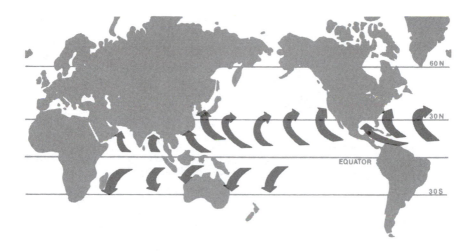

Figure 9-7. The paths of hurricanes take them northward in the northern hemisphere and southward in the southern hemisphere. As they move northward in the northern hemisphere they begin to be affected by the upper level westerlies and move more toward the east.

enced by the westerlies that cause them to curve and travel toward the northeast and east.

SOME FEATURES OF HURRICANES

Hurricanes are generated primarily during the fall season of the year in the Northern Hemisphere with peak activity during the month of September. Warm ocean temperatures are so important in their development that temperatures greater than 26°C are required for their formation (Figure 9-8).

Hurricanes do not have weather fronts associated with them as do midlatitude cyclones, but they are circular storms with low central pressure. The structure of a hurricane is quite different from a tornado or midlatitude cyclone; although in the developing stage, the tropical disturbance consists of an area of clouds with ascending air over the low pressure center. This occurs because the hurricane is a surface generated storm much like the dust devil. If you have traveled through the Desert Southwest in the summertime, you have, no doubt, seen dust devils in progress. Some of these are quite small and extend only a few hundred meters upward. They are surface generated as the ground is heated by the sun, and are characterized

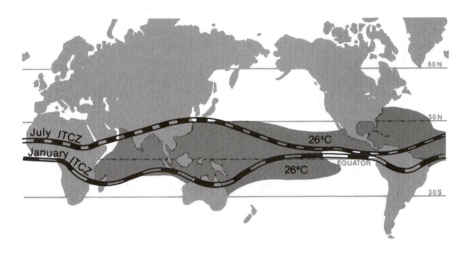

Figure 9-8. Hurricanes form only over areas where the ocean temperatures are quite warm, generally greater than 26°C and near the intertropical convergence zone (dashed lines).

by rotating air with a downdraft in the center. Hurricanes are also surface generated, but become a much larger storm, extending through most of the atmosphere to heights greater than 12 km. The eye of the hurricane is a region of light winds, few clouds, and low pressure. Figure 9-9 shows the eye of Hurricane Anita on September 1, 1977, as this hurricane was located just off the coast of Texas. The average width of the eye of a hurricane is about 40 km.

You might assume that because of the lower pressure the air would rise through the core of the hurricane. However, descending air occurs in the eye of the hurricane because the storm is surface generated with the centrifugal force throwing air outward from the core as it rises. The eye then contains only air coming downward to replace the air that is thrown outward by the vortex. The descending air is warmer than surrounding air and frequently causes cloud-free skies within the eye of the storm.

PRESSURE

The intensity of a hurricane is related to its central pressure. The lowest pressure ever recorded in a tropical cyclone was 877 millibars (25.9 in) during Hurricane Ida in the Philippines in 1958. One hurricane has reached the United States with a pressure less than 900 mb.

Figure 9-9. The eye of the hurricane is frequently clear or contains very thin cloud cover. This satellite photograph shows Hurricane Anita on September 1, 1977. (NOAA photograph.)

This was a hurricane that struck the Florida Keys in 1935 with a pressure of 892 mb (26.3 in). This pressure is much lower than the traveling frontal cyclones that come through midlatitude locations with accompanying winds and rain. Their central pressure seldom falls below 960 mb.

WIND SPEEDS

The wind speed in a typical hurricane is 175 km/h (105 mi/h) while the greatest that has been measured was about 350 km/h in Hurricane Camille in 1969. The amount and type of wind damage depends on the speed. Wind speeds of 60 km/h with gusts up to 70 km/h may damage TV antennas or awnings on windows. Higher speeds of 80 to 100 km/h cause roof damage to houses. Wind speeds greater than

120 km/h begin to produce structural damage to the roofs, eaves or other parts of houses. With wind speeds of 175 km/h and gusts slightly higher, major structural damage to buildings is common. As the wind speeds increase above 250 km/h buildings and cities may be reduced to rubble (Figure 9-10).

Measurements show that the highest winds are just outside the eye of the hurricane in a fairly narrow band (Figure 9-11). The speed decreases quite rapidly outward from this band of strongest winds. The wind velocity is very low in the eye, very strong just outside the eye and weakens again farther away from the eye.

SIZE

The size of a hurricane can be measured in two different ways. One method is to use the area within the last closed isobar surrounding the low pressure center as the outside boundary of the hurricane. On that basis, a composite of 100 storms gave an average size of 500 km

Figure 9-10. The strong winds associated with a hurricane are a very damaging component as shown here by the results of Hurricane Fredrick in September 1979. This marina on Dolphin Island, approximately 20 km south of Mobile, Alabama was destroyed. (US Coast Guard photograph.)

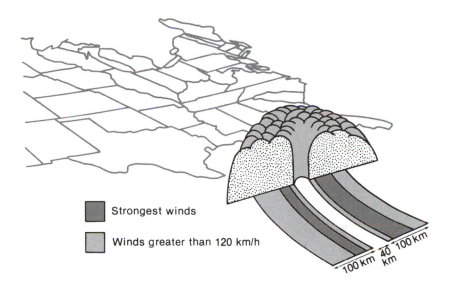

Strongest winds

Winds greater than 120 km/h

Figure 9-11. The location of the strongest winds and typical size of a hurricane are shown here. The band of strongest winds is just outside the eye of the hurricane. Winds are stronger on the right-hand side of the eye because of the addition of speed of travel of the hurricane.

in diameter for the hurricane. If you consider only the winds that are greater than hurricane velocity, 120 km/h, the average hurricane is about 250 km in diameter. This combination of size and intensity makes the hurricane a very destructive storm.

SPEED AND LIFETIME

The average speed of travel of a hurricane is 20 km/h, while the average lifetime is nine days. Thus the typical hurricane moves rather slowly, but lasts long enough to cover a considerable area.

RAINFALL

In addition to the intense winds generated by a hurricane, the amount of rainfall is a major problem. Many million dollars worth of damage can be done by rainfall amounts of 25 cm or less. An average hurricane produces a tremendous amount of water. A hurricane that struck Texas in 1921 brought 57 cm (23 in) of rainfall with it. Most cities do not have drainage facilities that can handle much more than 10 cm of rainfall per hour.

The rain in hurricanes is most intense in the spiraling bands around the eye. Thunderstorms are embedded in these spiraling bands and produce heavy rainfall. They also have lightning and thunder associated with them.

As a hurricane approaches land it brings warm, humid air that is usually sufficiently different from the air located over the land to create a weather front to add to the deluge of water (Figure 9-12).

DISSIPATION OVER LAND

Hurricanes dissipate very rapidly after landfall. The less humid air and rougher terrain drastically cut down the conversion of water vapor to cloud droplets with a reduction in the amount of latent heat to drive the hurricane. Thus, when they strike land they lose power very rapidly. The intensity of the hurricane drops by 50% by the time it is only 250 km inland. If the winds in the hurricane were 200 km/h when it struck the shore, they would have decreased to 100 km/h after moving 250 km inland. Coastal areas are therefore affected most by hurricanes.

The damage potential decreases with distance from the shore with 90% of the damage done within 100 km of the ocean, 50% a little farther inland, and no damage by the time the storm reaches as far

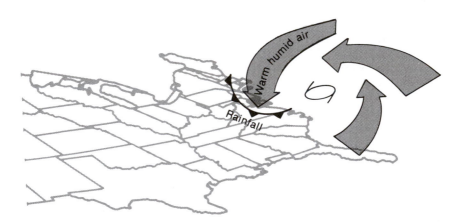

Figure 9-12. As a hurricane moves inland the strong circulation around it may form a warm front ahead of the leading edge of the hurricane. This adds additional precipitation and causes even greater damaging effects.

inland as Oklahoma or Arkansas for example. The damaged area from Hurricane Carla that struck the Gulf Coast in 1961 is shown in Figure 9-13.

The coastal regions that experienced inundation of water from Hurricane Carla are also shown. This factor causes much additional damage in low-lying areas. More damage typically occurs on the right-hand side of the path of the hurricane than on the left. If the hurricane is moving northward at a speed of 30 km/h, then the winds on the right-hand side will be 60 km/h faster than those on the left-hand side because the forward speed adds to the winds on the right-hand side and subtracts from those on the left-hand side. This effect can be seen in the damage path of Hurricane Carla. Captains of ships have used this information for decades, before the age of satellites, to avoid the strongest winds in a hurricane.

TORNADOES IN HURRICANES

Hurricanes add to their damage potential by producing tornadoes with even higher wind speeds. They are contained within the large thunderstorms that form the spiraling bands in advance of the eye of the hurricane. There is no preferred time of day for the formation of tornadoes within hurricanes. The rotation of the hurricane provides enough energy for individual thunderstorms to generate tornadoes without contributions from surface heating to help form severe thunderstorms.

Tornadoes with hurricanes are generally less severe than other tornadoes. Not only are tornadoes that are generated by hurricanes less intense, but they have shorter and narrower paths as well as shorter lifetimes. Other differences include the absence of the temperature inversion that is usually a factor in the generation of tornadoes in the Great Plains, no sharp vertical moisture stratification, and the absence of a negative lifted index.

A comparison of many tornadoes produced within hurricanes (Figure 9-14) shows that the forward right-hand quarter of the hurricane is the most likely location for the generation of tornadoes. They have also been observed in the left forward part of hurricanes, but are not as common there.

Numerous cases of tornadoes associated with hurricanes have been documented, although many more probably occur than can be re-

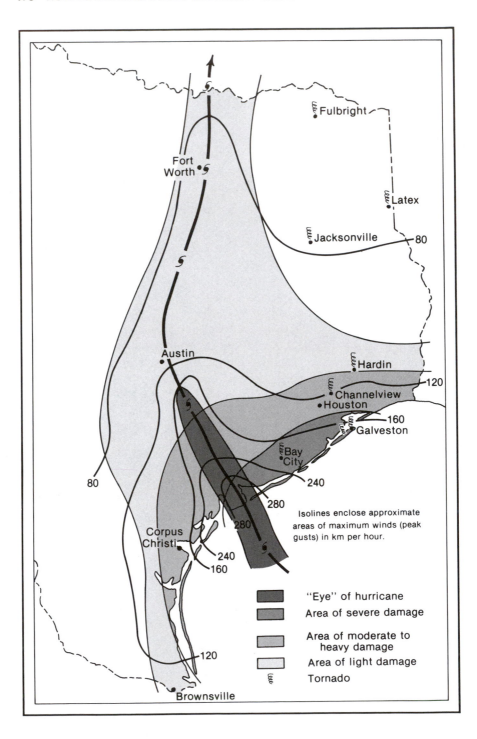

Isolines enclose approximate areas of maximum winds (peak gusts) in km per hour.

"Eye" of hurricane

Area of severe damage

Area of moderate to heavy damage

Area of light damage

Tornado

corded. Hurricane Carla in 1961 held the record for tornado production with 26 until Hurricane Beulah produced 115 confirmed tornadoes in Texas in 1967.

OCEAN SWELLS

Still another extremely damaging feature of a hurricane is the ocean
swell or storm surge, an increase in ocean level that it produces at
considerable distance from the storm. The high wind speeds within a
hurricane can generate waves to heights of 10% of the speed of the
winds in kilometers per hour for wave heights expressed in meters.
Such waves produce ocean swells characterized by undulating action
of the ocean in advance of the storm center. Ocean swells are produced by the hurricane winds and are propagated outward from the
area where they are produced. The winds on the right-hand side of
the hurricane generate swells that move in the same direction as the
hurricane. The strongest winds are on the right-hand side and produce the greatest swells which may reach the shore a day or more in
advance of the hurricane itself.

When the hurricane arrives it also brings a higher ocean level because of the inverted barometer principle. A hurricane with a pressure of 900 mb will have an increase in ocean level of more than 1 m
because of the lower atmospheric pressure.

Many of the rises in sea level for hurricanes in the Gulf of Mexico
have been more than 3 m above normal sea level. Hurricane Camille
in 1969 brought an increase in the ocean level of more than 7 m. If
you have driven along the coastal areas of the Gulf, you have seen
that a rise of this magnitude would inundate a large region.

In the Bay of Bengal off the coast of India, Pakistan, and Bangladesh, even higher waves are generated. The configuration of the
bay allows water to be driven into the narrowing portions to cause
large increases in ocean levels. Several tropical cyclones have produced increases in the ocean level within the Bay of Bengal of 12 m

Figure 9-13. The damages from Hurricane Carla are typical of those associated with a hurricane. Damages occur from inundations of coastal areas because of the high water levels.
Greater damage occurs on the right-hand side of the eye of the hurricane because of the strong
winds there. The area of greatest damage extends outward on each side of the eye and
decreases as the hurricane moves farther inland. (After Climatological Data.)

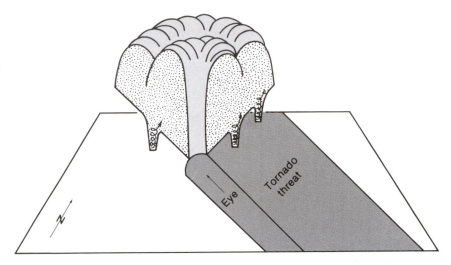

Figure 9-14. Most tornadoes associated with hurricanes are located in the leading right-hand quarter of the hurricane. The air drawn into the hurricane and its forward movement create a compression effect there and intensify thunderstorm activity.

or more.   Because of the dense population of these countries such tropical storms take a heavy toll of human life.

SUMMARY

Hurricanes are atmospheric vortices intermediate in size between frontal cyclones and tornadoes. Hurricanes originate as tropical disturbances over warm oceans within the tradewinds. The tropical disturbance intensifies into a tropical depression, then into a tropical storm, and finally becomes a hurricane as the winds increase to 120 km/h.

Hurricanes originate within the tradewinds near the intertropical convergence zone with the latent heat of condensation providing the energy for driving the storms. They form over warm oceans from 5 to 30° N and S latitude.

The structure of the hurricane includes an eye about 40 km in diameter with low pressure and descending air. The lowest recorded pressure of 877 mb was in Hurricane Ida in 1958. The typical hurricane has hurricane winds of at least 120 km/h over an area 250 km in diameter. Hurricanes typically travel at 20 km/h and last for 9 days.

They produce heavy rainfall and large ocean swells, but decrease in intensity quite rapidly after landfall.

Hurricanes may produce tornadoes within the leading rainbands. They are generally smaller than average but can be very damaging. Another damaging characteristic of hurricanes is the large ocean swells that they produce. These have been as great as 7 m in the Gulf of Mexico or 12 m in the Bay of Bengal.

10
Famous Hurricanes and Tropical Cyclones

As I read the headline, "Killer Camille claims 250th life," I wondered what kind of person could perpetuate such a crime against humanity. Surely the FBI and local police could have solved the case before it reached this magnitude. Where is the world headed? Will it soon be unsafe to walk down the street? And then I saw the rest of the story. Camille was no lady, she was a hurricane. . . .

HURRICANE NAMES

The practice of naming hurricanes after women was begun with the 1953 hurricane season. The name list for hurricanes in the Atlantic and Gulf Coast now has an international flavor because hurricanes are tracked by the public and the weather services of countries other than the United States. Names are selected from library sources and agreed upon during international meetings of the World Meteorological Organization by those nations involved. Male as well as female names were used for the first time in 1978 for Pacific hurricanes, and in 1979 for Atlantic hurricanes.

The National Hurricane Center near Miami, Fla., keeps constant watch on oceanic storm-breeding areas for tropical disturbances which may herald the formation of a hurricane. If a disturbance intensifies into a tropical storm, the Center gives it a name from the current list. Experience has shown that the use of short, distinctive given names is quicker and less subject to error in written and spoken

communications than more cumbersome latitude-longitude identification methods. The use of easily remembered names greatly
reduces confusion when two or more tropical storms occur at the
same time. Name lists for Atlantic hurricanes to be repeated every six years except for occasionally retired names are:

1991	1992	1993	1994	1995	1996
Ana	Andrew	Arlene	Alberto	Allison	Arthur
Bob	Bonnie	Bret	Beryl	Barry	Bertha
Claudette	Charley	Cindy	Chris	Chantal	Cesar
Danny	Danielle	Dennis	Debby	Dean	Diana
Erika	Earl	Emily	Ernesto	Erin	Edouard
Fabian	Frances	Floyd	Florence	Felix	Fran
Grace	Georges	Gert	Gilbert	Gabrielle	Gustav
Henri	Hermine	Harvey	Helene	Hugo	Horrtense
Isabel	Ivan	Irene	Issac	Iris	Isidore
Juan	Jeanne	Jose	Joan	Jerry	Josephine
Kate	Karl	Katrina	Keith	Karen	Klaus
Larry	Lisa	Lenny	Leslie	Luis	Lili
Mindy	Mitch	Maria	Michael	Marilyn	Marco
Nicholas	Nicole	Nate	Nadine	Noel	Nana
Odette	Otto	Ophelia	Oscar	Opal	Omar
Peter	Paula	Phillippe	Patty	Pablo	Paloma
Rose	Richard	Rita	Rafael	Roxanne	Rene
Sam	Shary	Stan	Sandy	Sebastien	Sally
Teresa	Tomas	Tammy	Tony	Tanya	Teddy
Victor	Virginie	Vince	Valerie	Van	Vicky
Wanda	Walter	Wilma	William	Wendy	Wilfred

EXCEPTIONAL TROPICAL STORMS

Twelve tropical cyclones are described individually in this chapter. Seven of these are "billion dollar hurricanes" based on the
amount of damage caused in the United States. Other storms
included are Carla, as an example of transformation of a hurricane into a frontal cyclone; Donna, which has the distinction of
affecting the whole East Coast from the tip of Florida to the tip
of Maine; Beulah since she generated a record number of tornadoes; the great Galveston storm of 1900 and the Bangladesh
storm which was the worst natural disaster.

Several of these hurricanes were modified into frontal cyclones as they moved across the United States. The paths of the hurricanes are shown in Figure 10-1.

The greatest loss of life from a hurricane striking the United States was in the Galveston hurricane of 1900. This great natural disaster was surpassed by the loss of life from the Bay of Bengal cyclone of 1970.

HURRICANE BETSY

Hurricane Betsy became the first of the three billion dollar hurricanes as it struck the Florida Keys on September 8, 1965, and Louisiana the next day. The tropical depression that was destined to become one of the most destructive storms in United States history was discovered on August 27 from satellite photographs and Navy reconnaissance aircraft. It was located at about 600 km east of Barbadoes. It reached hurricane intensity on August 29 and moved in a general northwestward direction. As it travelled some 500 km east of Florida it began to slow down on September 4 and gradually moved in a

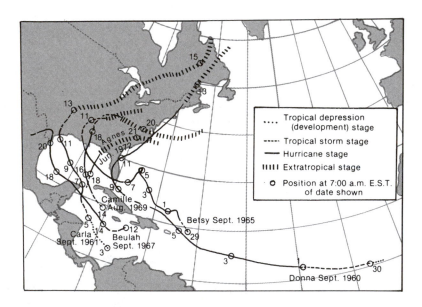

Figure 10-1. Paths of hurricanes Betsy, Camille, Agnes, Carla, Donna, and Beulah.

small clockwise loop as a large high pressure area over the eastern United States blocked its northward movement. By September 5 Betsy was travelling southward on a path that would take her around the tip of Florida and into the Gulf of Mexico. This was only the second hurricane in history to enter the Gulf of Mexico after moving southeastward.

Betsy brought a pressure of 952 mb within the 60 km eye with a band of winds that reached 230 km/h in southern Florida. The storm surge was as much as 3 m throughout the Florida Keys with a tide of 2 m along the Miami Beach oceanfront. Florida received $140 million damages from the storm.

Betsy's forward speed accelerated as it moved over the Gulf of Mexico before the eye crossed the southeastern Louisiana coast about 10:00 pm on September 9. Her lowest pressure was 948 mb with a storm surge of about 3 m.

The eye passed about 60 km southwest of New Orleans where the wind speed was estimated at 210 km/h. The estimated maximum winds throughout Louisiana are shown in Figure 10-2. Power and telephone lines were snarled along the Louisiana coastline and inland to north of Baton Rouge. Few residences or businesses in New Orleans escaped damages from the flooding and winds. Several communities near New Orleans were completely destroyed. In all 27,000 homes and more than 2,000 businesses were destroyed by this hurricane.

Although the total damage to property was extremely high ($1.4 billion) the death toll was well below record levels. The fatalities in Florida totaled only 4, with 9 additional deaths associated with small boats over adjacent waters. The deaths in Louisiana totaled 58 due to flooding or falling buildings. This number is amazingly low in comparison to the 250,000 people who fled from the coastal areas of Louisiana as Betsy approached.

At least six tornadoes were associated with hurricane Betsy. Three of these were in Florida, two in Alabama, and one in Mississippi.

HURRICANE CAMILLE

Hurricane Camille was the third of its season and has been called the greatest storm to strike the United States. On August 17, 1969, she slammed into the Gulf Coast and reminded man, once again, of his virtual helplessness in a full scale hurricane.

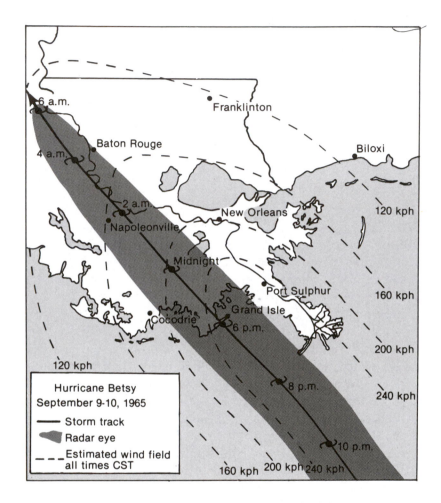

Figure 10-2. Hurricane Betsy brought winds of 250 km/h to the Louisiana coast. These were modified after landfall to less than 160 km/h six hours later as the eye passed Baton Rouge, as shown by the contours of wind speed. (Climatological Data, National Summary.)

Camille began as a small rainstorm and grew into a hurricane off the south coast of Cuba. Then, she roared into the Gulf of Mexico gathering fantastic power. The release of latent heat was very effective in driving the heat engine of hurricane Camille as currents of moisture rich air gave her incredible power; and she was fed by these air currents right up to the time she hit land.

An Air Force reconnaissance crew penetrated Camille's eye on August seventeenth. They radioed back a surface pressure of 901 mb

(26.61 in) and 230 km/h winds. The report prompted the Director of the National Hurricane Center, Robert Simpson, to state: "Never before has a populated area been threatened by a storm as extremely dangerous as Camille." A central pressure lower than Camille's had been recorded only once before in the North Atlantic—in the Labor Day Hurricane of 1935 near the Florida Keys. In size Hurricane Camille was small compared to other hurricanes. Hurricane winds (greater than 120 km/h) extended over an area only 150 km in diameter as Camille moved ashore. Only small storms such as Inez of 1966, had hurricane winds extending over smaller areas.

Noted for their devastating storm surges, Gulf Coast hurricanes bring with them large amounts of water. Camille generated ocean levels that were more than 5 m above mean sea level from Bay St. Louis to Biloxi. At Pass Christian, Mississippi the water level increased about 7 m (Figure 10-3). From southeastern Louisiana to Biloxi, Miss., almost total destruction was brought to waterfront areas from the combination of severe winds and devastating tides. Hurricane Camille became the most damaging storm in history with property damages of $1.42 billion. Cars and trucks were smashed like toys and giant freighters were tossed about and beached. One of those viewing the devastation described it as having the same appearance as if a giant bulldozer had leveled the whole coastal area.

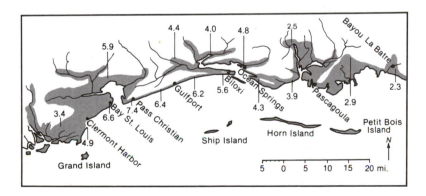

Figure 10-3. Storm surge and flooded areas as a result of Hurricane Camille. Water levels greater than 6 m above sea level were recorded from Bay St. Louis to Gulfport. This was the greatest storm surge ever recorded for the United States. (Climatological Data, National Summary.)

Hurricane Camille also induced tremendous amounts of rainfall. The Gulf Coast area received between 10 and 30 cm while the periphery of Camille received somewhat less. Most hurricanes do the majority of their damage along the coast where they hit. As Camille moved inland she was expected to die a normal death. However as the center of the remaining low pressure moved into Kentucky and the upper part of the storm interacted with the jet stream it turned abruptly toward the east and began dropping torrential rains. As Camille traveled over the Appalachian Mountains flash floods were common (Figure 10-4). Within a twenty-four hour period Massies Mill, Virginia, received an unbelievable 68 cm (27 in) of rain. On the 19th of August, Camille brought up to 61 cm (2 ft) of rain over the James River basin (Figure 10-5), flooded it, and almost doubled the death toll. Camille caused 144 deaths mostly from drowning along the Gulf Coast, while 114 deaths occurred in Virginia and West Virginia from flash floods. An incredibly large number of families (77,985) suffered losses from the hurricane.

Figure 10-4. Hurricane Camille produced disastrous flooding at the James River. The streets of Richmond, Virginia, were navigable only by boat as shown in this photograph. (Courtesy of Richmond Times Dispatch.)

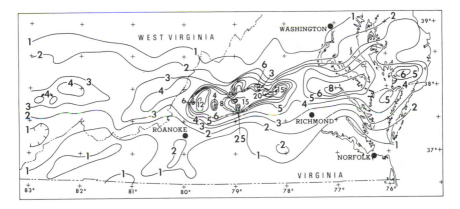

Figure 10-5. Rainfall distribution (inches) over Virginia as Hurricane Camille was transformed into an extratropical cyclone with a central updraft core. Some localized areas received even greater rainfall depths than those shown above. Most of Virginia received enough rainfall to produce severe flooding. (Climatological Data, National Summary.)

HURRICANE AGNES

The devastation of Hurricane Camille in 1969 was surpassed by hurricane Agnes in 1972. This was also a relatively small storm, but caused the greatest amount of damage (over $3 billion) of any storm in history. Like Camille, Hurricane Agnes struck the Gulf Coast and was downgraded from a hurricane. As it moved overland it interacted with the jet stream and again flooded many of the same rivers on the east coast after crossing the Appalachian Mountains. Florida, Maryland, New York, Pennsylvania, and Virginia were declared disaster areas. The number of people evacuated because of Hurricane Agnes was again, like Camille and Betsy, more than 250,000. Dikes broke in Richmond, Virginia. Harrisburg, Pennsylvania, was flooded by the Susquehanna River which rose to its highest level in almost two centuries of record keeping (Figure 10-6). Many houses were flooded, including the governor's mansion where the water rose to the first floor ceiling. The Potomac crested in Washington at 2 m above flood stage.

The total death toll in Hurricane Agnes was 117 with 48 of these in Pennsylvania. Agnes' heavy rains in Pennsylvania fell on wet ground. This increased the runoff into the streams and contributed to the flood problem. Rainfall amounts over the entire central part of Pensylvania ranged from 20 to 30 cm.

Figure 10-6. Hurricane Agnes produced severe flooding in the eastern United States. The Susquehanna rages through the downtown business section of Wilkes-Barre, Pennsylvania causing millions of dollars of damage. At center smoke rises from a burning building as fire fighters train hoses on the blaze. (Official US Coast Guard Photograph.)

Agnes spawned 15 confirmed tornadoes in Florida on the 18th and 19th of June. Total damage from these tornadoes was $4.5 million and more than 40 people were injured by them.

It is noteworthy that a major part of the damages caused by Hurricane Agnes occurred as it moved inland and was transformed into an extratropical cyclone. Its interaction with the jet stream is shown in Figure 10-7. As it encountered these westerly winds they became a part of the cyclonic circulation to give it the same features as a typical frontal cyclone. This allowed the storm to produce heavy rains after it crossed the Appalachian Mountains. Thus, the transformed storm was almost as damaging as the hurricane stage.

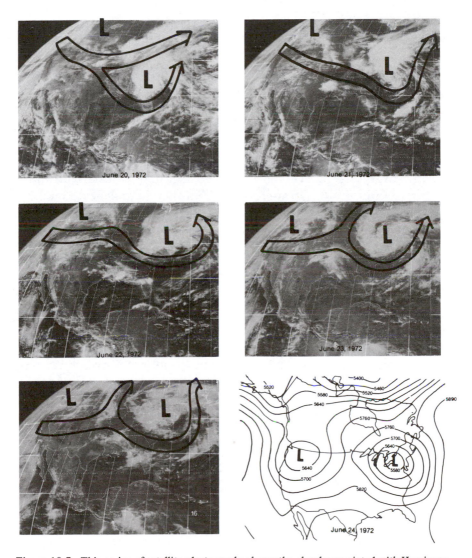

Figure 10-7. This series of satellite photographs shows the clouds associated with Hurricane Agnes after it struck the gulf coast and was modified into a frontal cyclone. The arrow shows the position of the upper air current and the 500 mb map is also shown for June 24th. Major flooding was produced in the Eastern United States by this storm.

HURRICANE FREDERIC

A tropical storm far out in the Atlantic Ocean reached hurricane intensity on September 1, 1979 and was named Hurricane Frederic. After weakening and passing over Cuba only one week after Hurricane David, Frederic began to regain strength in the Gulf of Mexico. He headed directly for Mobile, Alabama and gale warnings went into effect from Grand Isle, LA, to Panama City, FL, on September 11. Frederic struck land with a central pressure of 946 mb and winds of 213 km/h (132 mph). A storm surge of 4 m over Gulf Shores, Alabama destroyed much of the island. A similar surge over Dauphin Island destroyed the causeway leading to the island. Frederic killed five people and produced a damage total of $2.3 billion.

HURRICANE ALICIA

On August 15, 1983 a tropical depression over the north central Gulf of Mexico strengthened to a tropical storm and it was named Alicia. The storm strengthened rapidly as it headed toward the coast. Alicia made landfall about 25 mi southwest of Galveston on August 18. Ocean swells of about 4 m were observed in Galveston Bay. Sustained winds were estimated at 185 km/h (115 mph). Rainfall amounts approached 11 in east of Houston and 23 tornadoes were reported ahead of the hurricane.

The total damage from Alicia was estimated to be about $2 billion, which made it the costliest hurricane in Texas history. The number of insurance claims filed was 275,000 which provides an indication of the number of people who lost property from this hurricane. Twenty one persons were killed by the storm.

HURRICANE ELENA

On August 29, 1985 a tropical storm became Hurricane Elena as it moved into the Gulf of Mexico from its location over Cuba. As Hurricane Elena approached the Gulf coast a frontal trough caused it to make an anticyclonic loop off the west coast of Florida. Its erratic course caused great difficulty in predicting its future position, but finally took it ashore near Biloxi, Mississippi on September 2. Maximum sustained winds were 169 km/h (105 mph) with gusts to 135 mph. Ocean swells of 3 m above normal were observed at Apalachicola, Florida.

More than one million people were evacuated from coastal areas during Elena's erratic approach, although only four people were killed by the storm. Damages caused by Hurricane Elena pushed her into the billion dollar category as the total reached $1.25 billion.

When the storm was over, the Civil Defense Director for the Gulfport area commented, "Elena was the most aggrevating storm I have ever been associated with in 20 years. She taunted us, she teased us, and then she hit us."

HURRICANE HUGO

Satellite photographs showed a cluster of thunderstorms on September 9, 1989 moving off the coast of Africa. A tropical depression formed southeast of the Cape Verde Islands and moved westward to become Hurricane Hugo on September 13 about 1250 miles east of the Leeward Islands. Hugo's estimated surface wind speed on September 15 was 258 km/h (160 mph) and pressure was 918 mb, earning the hurricane a category five rating (most intense rating from the Saffir/Simpson Hurricane Scale). Hurricane Hugo struck Guadeloupe Island on September 17 and St. Croix one day later with winds of 140 mph. The hurricane then accelerated and struck the islands of Vieques and Puerto Rico before heading for South Carolina.

Hugo made landfall on the 21st at Sullivan Island near Charleston with winds estimated at 140 mph. Storm tides along the South Carolina coast ranged from about 3 m in the Charleston-Folly Bay to about 6 m in the south end of Bulls Bay. A 150-mile wide swath of 3 to 8 in of rain spread inland across South Carolina and western North Carolina.

The strength and long life of Hurricane Hugo resulted in an intense amount of damage. Total damage from Hugo was estimated to be $7 billion making this hurricane the costliest in history. Hurricane Hugo was responsible for 49 deaths some of which occurred outside the United States.

HURRICANE CARLA

A less destructive hurricane was Hurricane Carla as it hit Texas in 1961 and caused damages amounting to $200 million. It was a strong hurricane that also changed into a frontal cyclone as it moved inland and interacted with the jet stream. It became a typical traveling low pressure system of the type that originates over continents except that it was smaller and had more rain associated with it. It dumped 30 cm (12 in) of rain on Kansas City, for example, although the wind velocities were not high enough to cause damages very far inland. The amount of rain was similar to the amount that caused flash flooding at the Plaza Shopping Center in Kansas City in 1977 and resulted in the death of 25 people, but the rains from the remnants of Hurricane Carla were spread over 2 days.

Hurricane Carla remained at hurricane intensity from the 3rd to the 12th of September. Hurricane Carla had a pressure of 960 mb with several closed isobars just before it hit the Texas coast. It brought hurricane winds over an area 300 km in diameter with strongest winds of about 130 km/h. It was modified very rapidly after landfall so that by September 12 the isobars showed less packing and the winds had decreased to 40 km/h with a surface pressure of 980 mb as it moved farther inland over the central United States. Indeed Hurricane Carla had become a frontal cyclone (Figure 10-8) as it interacted with the jet stream and the circulation around the surface low created a cold and warm front from a previous stationary front. Perhaps it would be more accurate to call the storm a "hurriclone" as such hybrid storms maintain a greater capability for producing precipitation than an ordinary frontal cyclone. By September 14 the center of the storm was over Minnesota where it continued to produce precipitation near the center of the storm and along the fronts.

Loss of life in Carla totaled 46 with 34 deaths in Texas, including 8 in a Galveston tornado, 6 died in associated tornadoes in Louisiana, 5 in flash floods in Kansas, and 1 in Missouri. Carla produced at least 26 confirmed tornadoes that also contributed substantially to the amount of damage.

HURRICANE DONNA

On September 2, 1960, Hurricane Donna was spotted 1200 miles east—southeast of San Juan, Puerto Rico, from ship reports and Navy

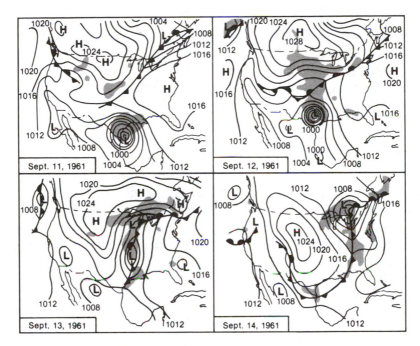

Figure 10-8. This series of synoptic charts for September 11 to 14, 1961 shows the transformation of Hurricane Carla into a typical frontal cyclone except for the unusually heavy rainfall associated with the transformed storm.

reconnaissance aircraft. A Navy aircraft was sent to Hurricane Donna and found an intense storm on September 2nd with winds estimated at 240 km/h. More than a week later, on September 10, Hurricane Donna struck the Florida Keys with winds of this same speed. As shown in Figure 10-1 Donna passed over much of Florida as it turned toward the north. Hurricane Donna did not die over Florida but only experienced a decrease in wind speeds down to about 120 km/h with an enlargement of the eye as it passed over central Florida. As the storm moved over the ocean just north of Daytona Beach on September 11, the hurricane structure remained intact and began to intensify. The winds increased to 200 km/h.

As Hurricane Donna moved northeastward along the coast the large eye was a continuing feature of this hurricane. Its path took it inland again on September 11 in North Carolina. As it moved over such cities as Swansboro and Elizabeth City the winds were about 120 km/h with an hours lull as the eye passed over with clear skies and light winds. Its northeastward path took it over the Atlantic again where it was not able to reintensify to its previous strength although the winds were above hurricane strength by the time it struck Long Island with winds of 150 km/h. Winds reached 80 km/h in New York City on September 12.

Hurricane Donna continued to move northward and affected Portland, Maine with 100 km/h winds before finally being downgraded to the extratropical cyclone stage over central Maine. Hurricane Donna thus has the distinction of bringing severe weather to the whole eastern coastline of the United States from Miami, Florida, to Caribou, Maine.

Damages in Florida amounted to about $300 million with 13 deaths. Georgia and South Carolina reported almost $1 million damages with no deaths. North Carolina and Virginia received $56 million damages with 11 deaths. Maryland, Delaware, Pennsylvania, and New Jersey sustained several million dollars worth of damages with 12 fatalities. New York and New England sustained at least $15 million damages with 6 deaths.

HURRICANE BEULAH

Hurricane Beulah of September 1967 has the distinction of producing more tornadoes than any other hurricane. It should also be mentioned that the name Beulah was used previously for a hurricane in August 1963. Although the first Beulah did not strike land, it is also referred to occasionally since it was one of the few hurricanes seeded for purposes of modification.

The weak tropical depression that was to become hurricane Beulah in 1967 formed on September 5 east of the Windward Islands. Hurricane winds were developed by the next day and reached 230 km/h prior to touching the southern coast of the Dominican Republic on September 11. Beulah caused considerable damage among the Caribbean Islands before reaching the Gulf of Mexico where she could regain the strength lost because of the rougher terrain of the Islands and loss of warm, humid air from over the ocean. Before landfall on September 20 just east of Brownsville, Texas her pressure was 923 mb with winds of 200 km/h.

Property and crop losses in Texas were about $200 million. Fifteen persons died in Texas from Hurricane Beulah. Five of these were killed by tornadoes. At least 115 tornadoes were spawned by Beulah

(Figure 10-9). Most of the tornadoes were small and formed over rural areas where they caused only minor damage. There were exceptions, such as the tornado that struck Palacios on the morning of the 20th. It picked up ten people and carried them up into the tornado. As they were dropped in a field four of them were killed and the others were injured. A tornado also struck Burnet and moved across the city with resulting damages of $100,000. Damages of at least this magnitude were experienced in several other communities.

The path of Hurricane Beulah contributed to the large number of tornadoes that were formed. As Beulah approached Texas and turned toward the west the strong east winds ahead of the eye, carrying tons of moisture, interacted with and then replaced the general westerly winds over Texas (Figure 10-9). Such opposing air currents are sufficient to generate tornadoes as has been demonstrated in the laboratory. Thus, Beulah became the best tornado generator in the history of hurricanes. She was surpassed only by the April 3 and 4, 1974, frontal cyclone that generated 148 large tornadoes.

GALVESTON HURRICANE OF 1900

Charts of the Gulf of Mexico made long ago show three great oak trees as the landmarks of Galveston, Texas. For three centuries storms in the area battered those big trees, harming them very little. High winds and tides caused some destruction, but never enough to slow Galveston's development—or uproot the sturdy oaks. Then on September 8, 1900, a hurricane swept over the island causing catastrophe and felling the great oaks. The death toll of more than 6000 places this storm as the greatest natural disaster in this nation's history.

It was not until the evening of September 7 that a storm warning flag was put up. This was the only hint, besides a very small article in the "Galveston News," that a storm was coming. The "brick dust" sky which traditionally heralds the approach of a hurricane was also missing that day. Everyone knew the city was almost totally helpless against the sea. There was no high ground, as the highest point of the city was less than 3 m above sea level.

Isaac Cline, head of the U.S. Weather Bureau office at Galveston, had started to monitor the storm at 5 am on Saturday, September 8, not knowing that the hurricane would zero in on Galveston. There had been light rain early in the morning then it started to pour. By

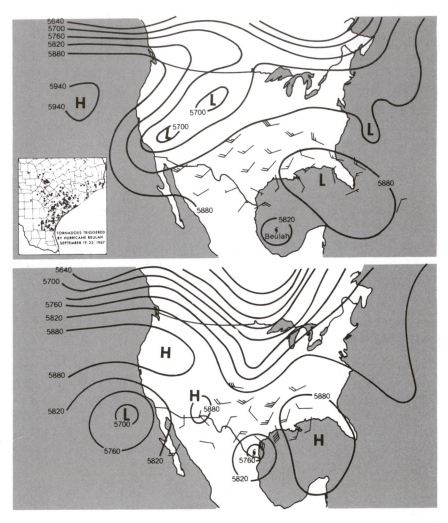

Figure 10-9. As Hurricane Beulah approached Texas on September 19, 1967 (top), the airflow at 500 mb was from the southwest over much of Texas. As the storm reached southern Texas, two days later (bottom) these winds were replaced by easterly winds north of the hurricane center. This wind shear may have played a role in the formation of many tornadoes.

3 pm, over half the city was under water with many people trapped in their low-lying houses. The exact speed of the wind during the fifteen hours that it battered Galveston is not known since the Weather Bureau's anemometer blew away after registering 84 mi/h (140 km/h) at 5:15 pm. It is estimated that the winds reached 200 km/h blowing masses of salt water across the island.

The four bridges to the mainland had washed out early, and all contact with the rest of the world was cut off by midafternoon. The city's water and electric plants ceased to function. By late afternoon there was water over every foot of the island, with the center, the highest part, under more than 1 m of debris-filled water. At 8:10 pm, the barometer stood at 28.53 in, the lowest it had ever been in Galveston. Many people sought sanctuary in churches and hospitals that collapsed, killing almost everyone inside. At about 10 pm, the raging winds abated, and for a few moments the survivors had a tremendous sense of relief, but they were to suffer more. As the wind, which had carried huge quantities of water onto the island, died, a massive surge of waves headed seaward, creating more destruction and drowning many people who survived the hurricane's original assault.

Over 1500 acres of the city had been devastated along with well over 3800 houses. There was an estimated $30 million of property damages. The death toll is listed at 6,000, although discrepancies place the estimated total at over 7,200, based on evidence that many neighborhoods were destroyed and left no friends who could furnish names. Added to this fact is that many strangers and dock workers were in the city with no local ties. It was also impossible to get the precise number of people in some families.

The weeks following the hurricane were grim times, but the Galvestonians would not consider outsiders' advice to abandon the island. Following an incredibly successful long-term rebuilding program, Galveston incorporated such measures as raising the entire grade of the city from 1 to 5 m above the old level, constructing a seawall along the Gulf (Figure 10-10), and connecting the island and mainland by a fine trafficway. By 1905 the seawall was 5 km long and was 6 m above the sea level while extending 3 m above the streets. The seawall was eventually extended to its present 17 km length.

Proof that the wall provided safety was realized as soon as 1915 when a great hurricane as strong as the one in 1900 killed over 250 in two states but killed only eight in Galveston and flooded the city

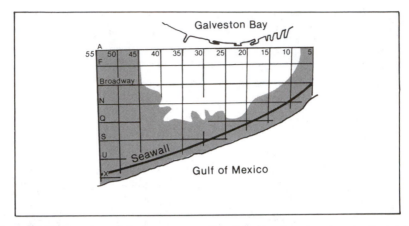

Figure 10-10. Location of the present seawall in Galveston in comparison to the total de-struction (shaded area) from the hurricane of 1900. Property south of the seawall (black line)was completely washed away. (After *Weatherwise*, 32:4, 1979.)

with only 1.5 m of water. The new trafficway went down, all except one section, which held a stalled two-car trolley filled with frightened passengers who survived. The improvements have provided protection several times since, particularly during Hurricane Carla in September 1961. Thus, Galveston remains a beautiful commercial and pleasure port.

BAY OF BENGAL CYCLONE OF 1970

For Bangladesh, November has always meant cyclone weather. This country of only 143,000 km² contains more than 78 million people. Most of the country is less than 16 m above sea level and borders the Bay of Bengal with lowlands of the Ganges Delta. This area faces storms that periodically take a high toll of lives. On November 12-13, 1970, in East Pakistan (now Bangladesh), a cyclone brought winds of 250 km/h. The winds of the storm drove the confined waters in the Bay of Bengal against the shore in waves of 7 m height. The resulting loss of human life was estimated to be at least 300,000.

On the night of November 12 there was a full moon and the tides were at their highest just before the cyclone ripped into the island and coastline. A United States satellite report (Figure 10-11) had been relayed to the Pakistanis, but their own meteorological system

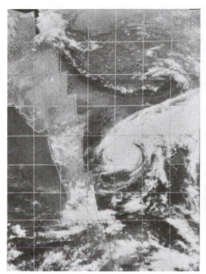

Figure 10-11. These satellite photographs for November 10, 1970 (left), and November 11, 1970, show the cyclone that moved into the Bay of Bengal with very devastating results. More than 300,000 people were killed by this tropical cyclone. (Courtesy of National Environmental Satellite Service, NOAA.)

had already predicted when the storm would strike. Warnings were broadcast, but there were not enough receiving sets, and the warnings said nothing about huge waves. Even if they had, it would not have altered the outcome, as most of the affected area was little more than a meter above sea level. When the big waves rolled in, there was no safe retreat for people as they drowned along with their doomed livestock. Those who did find temporary safety on rooftops saw the big waves approach with apprehension and were sure it was the end of the world.

Everywhere the rich rice fields were ruined. Drinking water was scarce. Food and medical supplies were not available. In a few days, there were outbreaks of cholera to claim more lives.

Immediately after the tragedy the United States pledged $10 million for relief. The People's Republic of China pledged $1.4 million. Britain dispatched her Royal Marines, a helicopter carrier, food, and medical supplies. India, too, sent money and food. Unfortunately, though there was no shortage of planes bringing in food and medical supplies, they did not all reach the people who needed them. There were too few airfields and most of these were submerged. Trucks were abundant, but there were no roads for them to use. Much of the incoming supplies ended up on docks and in Dacca's warehouses and a high percentage found their way into the black market.

This storm may not be the last to cause such a high death toll because of the combination of circumstances, including lowlands, high population density and the narrowing shape of the Bay of Bengal. Other storms in the Bay of Bengal have killed more than 100,000 people.

It is perhaps noteworthy in this regard that the loss of life from a single hurricane in the United States could be many thousands without the warning and evacuation procedures that are usually effective.

SUMMARY

Each hurricane in the Atlantic since 1953 has been given a name from a list prepared in advance. Names are repeated every six years except for those hurricanes that achieve some degree of fame. Hurricane Betsy brought intense winds to a major city, New Orleans, and produced heavy damage over a small area. Camille and Agnes both struck the Gulf Coast as intense hurricanes, then continued to produce heavy rainfall as they became extratropical cyclones. Camille had the strongest winds of any storm with gusts to 205 mph. Hurricane Hugo surpassed Agnes and caused the greatest amount of damage of any hurricane to strike the United States. Other billion dollar hurricanes were Betsy, Camille, Frederic, Alicia and Elena.

Hurricane Donna produced strong winds and precipitation along its path from the tip of Florida to Maine as it struck land for short distances then passed over water where it intensified before striking land again. Hurricane Beulah generated more than one hundred tornadoes to surpass the previous record of 26.

The Galveston hurricane of 1900 killed more than 6,000 people making it the greatest natural disaster in the United States, while the Bay of Bengal cyclone of 1970 killed more than 300,000 people in Bangladesh.

11
Prediction and
Modification of Hurricanes

According to the computer drawn map we would be entering the outer rainbands shortly. Verification came rapidly in the form of extremely turbulent air currents, heavy rain, and electrical displays that surrounded and buffeted our reinforced and specially equipped aircraft. Our carefully prepared statistical procedures dictated that we begin the opening of envelopes to direct our research activities in unison with pressure, temperature, and wind measurements. I could hardly believe that we were actually going to fly through the strongest part of this hurricane. As the 100 km/h winds engulfed us we read the first instructions: Seed outer rainbands, do not seed eyewall. We hurriedly read the second randomly selected set of instructions: Seed with silver iodide, do not use dry ice. Even as we pulled the appropriate levers and recorded numbers it was impossible to miss the sudden change in the outside environment as we broke into clear air and calm winds. There was no mistaking the circular wall of clouds that now surrounded us as we flew across the eye of the hurricane and prepared again for the fury awaiting us along any escape route we might choose. . . .

PREDICTION OF HURRICANES

There are no atmospheric conditions that we can measure and combine to predict where a hurricane will develop as we do for severe thunderstorms. Therefore hurricane prediction currently involves only the specification of the path the hurricane will follow after it is formed.

Satellites are very helpful in locating developing hurricanes and in charting their path after they have formed. The National Hurricane Center in Miami, Florida, has the responsibility for predicting the path of hurricanes after they have developed.

Although a hurricane may cover most of the Gulf of Mexico (Figure 11-1), the path of the eye is very important since the strongest winds are located near the surrounding eyewall.

PATHS OF HURRICANES

The typical movement of Atlantic hurricanes is initial travel from the southeast (Figure 11-2). As they move into more northerly latitudes,

Figure 11-1. This satellite photograph of Hurricane Allen shows detail of the cloud structure as this hurricane filled much of the Gulf of Mexico. This photograph was taken on August 8, 1980, and has 1/2 mile visual resolution. (Courtesy of National Environmental Satellite Service, NOAA.)

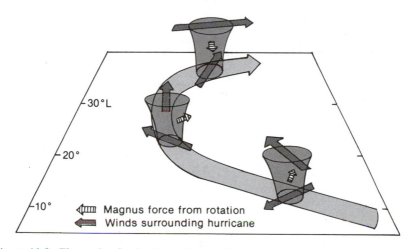

30°L

20°

10°

⬅🎏 Magnus force from rotation
⬅🎏 Winds surrounding hurricane

Figure 11-2. The path of a hurricane is related to the winds surrounding the hurricane including those at low and upper levels. The rotation of the hurricane also causes a force to the right of the path of motion. Various combinations of these forces cause different hurricane paths.

they generally start to curve with a more northward movement, followed by movement from the southwest. As hurricanes develop in the warm tropical waters of the Atlantic Ocean they generally take a curved path as shown. They may travel westward for a week, thus endangering the eastern coast of the United States even if they originate several hundred kilometers away. If a hurricane develops in the Gulf of Mexico it will almost certainly hit the Gulf Coast.

Some hurricane paths that were atypical are shown in Figure 11-3. A hurricane may suddenly change directions as it follows a looping path. This may cause it to strike land when its previous path would have kept it over the ocean. Hurricane Betsy was of this type. She was generated out in the Atlantic and maintained hurricane winds for over nine days before she struck the Gulf Coast. It first appeared that Betsy would reach the Eastern United States, but a looping path took her around the tip of Florida and into the Gulf instead. Such hurricanes that describe a looping pattern as they move northward make it almost impossible to predict their future movement while they travel in a circle.

The distribution of high and low pressure affects the paths of hurricanes as shown in Figure 11-4. Hurricanes will change their course to avoid either a high or low pressure system in midlatitudes. You might think that low pressure areas would attract a hurricane but they

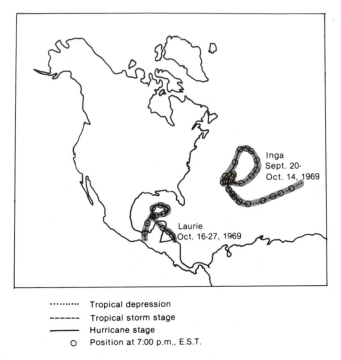

```
·········· Tropical depression
------   Tropical storm stage
───────  Hurricane stage
   O     Position at 7:00 p.m., E.S.T.
```

Figure 11-3. Examples of hurricanes with more unpredictable paths are shown for two hurricanes that occurred in 1969.

will go around either a midlatitude cyclone or high pressure system. Hurricanes will travel either eastward or westward to avoid a midlatitude cyclone or anticyclone located to their north and do not ordinarily combine with either because their structure is so different.

HURRICANE PATH PREDICTIONS

About 70% of the hurricanes behave in typical fashion by traveling in paths that are predictable. It is the other 30%, which includes hurricanes such as Betsy, that make predictions of future paths of hurricanes difficult. Another difficulty in providing precise warnings to people living in coastal areas arises from hurricanes such as Agnes and Camille since they only developed strong winds within the Gulf of Mexico shortly before landfall.

Most people who have experienced a hurricane have been in its fringe area, since the most intense winds are confined to a fairly small

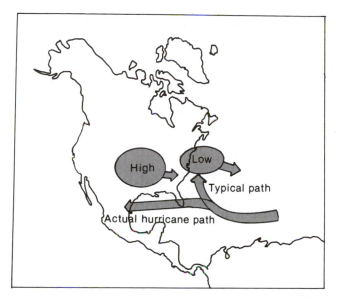

Figure 11-4. The path of a particular hurricane is also influenced by the location of travelling high and low pressure centers in midlatitude. Hurricanes tend to avoid either high or low pressure centers.

area. Therefore they may have a false impression of the severity of a major hurricane. This impression is reinforced by weaker hurricanes that strike land. This sets up situations such as that when 25 people held a hurricane party as Camille blasted the Richelieu Apartments in Pass Christian, Mississippi. Only two of them survived; one of them, a boy, floated out the window on a mattress. Both he and the other survivor were severely injured by the high winds, waves, and debris.

The National Hurricane Center in Miami uses four different track prediction models in an effort to pinpoint the location of landfall for hurricanes reaching the United States (Figure 11-5). One of these is purely dynamic; the others are statistical or climatological. If the location of landfall can be predicted, it makes a tremendous difference in warning people of the approaching danger. The accuracy of predicting the exact location of the center of the hurricane at landfall is only about 175 km. But this distance may mean that one particular city is affected instead of another. Timely warning is very important in evacuating a large city in the path of a major hurricane. None of the techniques predict the hurricane path perfectly. This is the reason

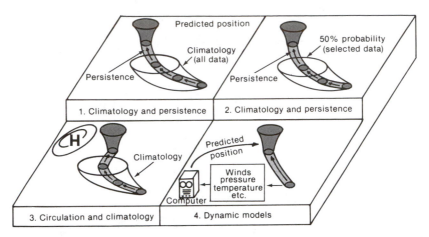

Figure 11-5. Hurricane track prediction techniques include the four different methods illustrated above. These are based on climatology, persistence, circulation, and dynamic models.

for having different prediction models to pinpoint as specifically as possible where the hurricane is going to reach land.

The first hurricane path prediction model used by the National Hurricane Center is based on climatology and persistence. Persistence is simply the behavior of the storm in the past. This includes speed of travel and its direction during the past. This direction and speed is then used to project the hurricane speed and direction for the next day. The climatology of previous hurricanes including the speed and direction of all other hurricanes is added to form the first technique for predicting the movement of a particular hurricane. This gives a prediction for the location of the storm 12 hours or 1 day in advance.

The second method used at the National Hurricane Center is similar to the climatology and persistence technique except that instead of using data on all previous hurricanes for the climatology of hurricanes, only data near the particular location of the current hurricane are used. Then the projection is based on hurricanes located in the same area combined with the persistence of the current hurricane.

The third technique used to predict the path of a hurricane is based on circulation and climatology. The climatology of past hurricanes is combined with such factors as the surrounding pressure systems, upper air flow patterns, direction of the trade winds, and other information.

The fourth method consists of a dynamic model. A computer processes equations that require information concerning winds at the surface, winds at 16 km and several levels in between. Thus data on the current atmospheric conditions are used to compute the location of the center of lowest pressure (hurricane center) at a later time.

None of these hurricane path prediction methods are consistently more accurate than the others. All are used for each hurricane that forms. The results of one method may be considered more reliable than the others for a particular hurricane from past experience, but no one method is better under all conditions. Figure 11-6 shows the predicted path and affected coastal area for a particular hurricane.

REMOTE SENSING OF HURRICANES

Although we cannot presently forecast areas where hurricanes are likely to develop based on atmospheric conditions, we can observe them soon after they form by satellite photography, since satellites take pictures of the whole earth every 15 minutes. Satellite photographs of many past hurricanes have been catalogued to provide a basis for remote sensing of hurricanes (Figure 11-7). Alternatives to remote sensing are to fly an airplane through the hurricane. This has been done in the past and is still a part of the activities of Air Weather Squadron of the Air Force. The National Hurricane Center in Miami, Florida, is responsible for providing information on the intensity as well as tracking and predicting the point of landfall for hurricanes near the United States.

Measurements of the lowest pressure and highest wind speeds in various hurricanes have been combined with satellite photographs for remote sensing purposes. The nature of the eye and surrounding cloud bands are used to make predictions about the intensity of hurricanes. If the cloud bands completely enclose the eye, it indicates a stronger storm. The width of the bands also give some indication of the storms intensity. If the clouds surrounding the eye are very irregular a weak storm is indicated. If the eye is round and large a stronger storm is indicated. The type of eye is also a factor. If a small eye has circular bands surrounding it, then the storm is more intense than if the bands are not closed around it. The nature of the eye and the banding features are used as criteria for assigning numbers indicative of a particular storm's intensity.

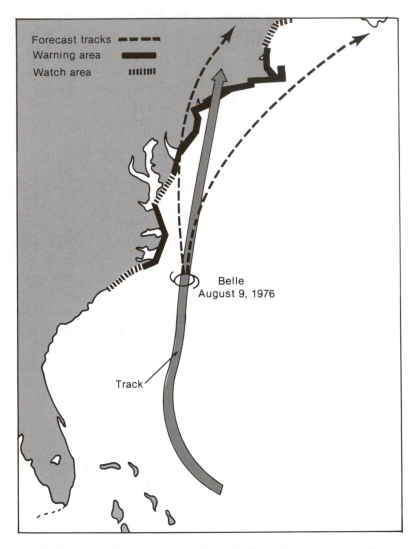

Figure 11-6. Forecasting the track of a particular hurricane is very important since several hundred thousand people must be evacuated from the path of a typical hurricane. The actual path and the forecast paths of Hurricane Bell are shown above, along with the coastal area affected by hurricane warnings and watches.

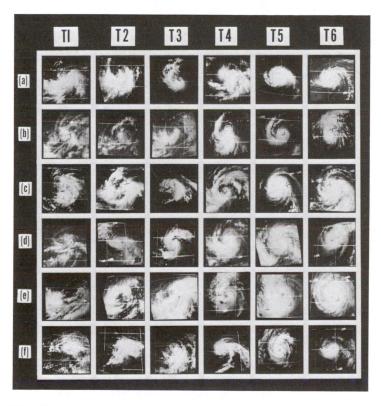

Figure 11-7. This series of satellite photographs shows 6 different hurricanes (A–F) as they developed in intensity and were assigned a classification number from T1 to T6. The related pressure and winds for these various intensity numbers can be obtained from Table 11-1. Such a catalog of hurricane photographs may be useful in future remote sensing techniques. (After Dvorak, 1973.)

Figure 11-7 shows some of the past satellite photographs of hurricanes with assigned numbers T1 to T6 on the basis of their appearance. This can then be related to characteristics of the hurricane as shown in Table 11-1. A number of T1 corresponds to wind speeds of only 46 km/h, while a number of T6 corresponds to wind speeds of 213 km/h and a pressure of 948 mb in the Atlantic (942 in the Pacific). This is a promising remote sensing technique using satellite photographs to gain information about hurricanes. Although it is not perfectly accurate, it is an example of a better approach than obtaining direct measurements through an intense hurricane.

Table 11-1. The empirical relationship between the T-number, the maximum wind speed (MWS), and the minimum sea level pressure (MSLP). The relationship between cloud features and the isotach pattern is as described in Dvorak (1973).

T NUMBER	MWS (km/h)	MSLP (ATLANTIC) (mb)	MSLP (NW PACIFIC) (mb)
1	46	1013	1007
1.5	46	1012	1006
2	56	1009	1003
2.5	65	1005	999
3	83	1000	994
3.5	102	994	988
4	120	987	981
4.5	143	979	973
5	167	970	964
5.5	189	960	954
6	213	948	942
6.5	235	935	929
7	259	921	915
7.5	287	906	900
8	315	890	884

SEEDING OF HURRICANES

The winds and water accompanying a hurricane are so damaging that considerable interest exists in modifying these features. Project Stormfury is a national effort to seed hurricanes and measure any resulting modifications. Seeding a hurricane consists of adding silver iodide or dry ice to it by means of an airplane. Dry ice crystals have a very low temperature of $-78°C$ and are capable of generating many ice crystals in a supercooled atmosphere. Silver iodide is a substance that has crystals so similar in shape to the ice particles that it also causes the generation of many ice crystals.

Hurricanes are seeded by adding the seeding agents to the hurricane in particular locations (Figure 11-8). Two different locations are used for seeding hurricanes. One train of thought is that seeding is best accomplished in the area close to the eye of the hurricane. The idea is to convert more of the water vapor to ice crystals to widen the eye and weaken the storm because of the principle of conservation of angular momentum as previously discussed. This principle is applicable to hurricanes as can be seen from the fact that the most damaging

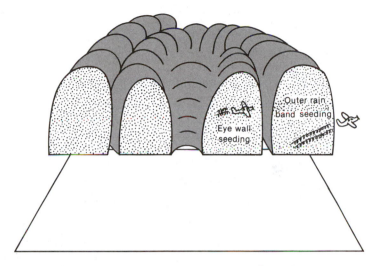

Figure 11-8. Hurricanes may be seeded near the eye wall to affect the pressure–wind relationship, or in outer rain bands to extract moisture from the winds before they get into the inner rain band.

hurricanes are generally the smaller ones like Camille and Agnes. If we can seed a hurricane and cause the diameter to increase, we would decrease the power of the winds.

The other train of thought is to seed several kilometers outward from the most intense winds to increase the latent heat and rainfall from the warm moist air traveling inward to feed the clouds around the eye wall. If these wind currents can be seeded and converted into large updrafts this could cut down on the amount of moisture that would be carried into the most intense region of the hurricane. This should reduce the most damaging winds near the eye.

Four hurricanes have already been seeded (Figure 11-9). The first was Hurricane Esther in 1961. Following this seeding measurements indicated that more ice crystals were produced. The other observed characteristic was that it began to describe a looping path and struck the United States. There is no direct evidence that the seeding was the cause of the change in direction since none of the others that have been seeded showed a similar change in direction. Since it struck the United States it resulted in new rules for seeding hurricanes. They must now be a much greater distance away from land when they are seeded.

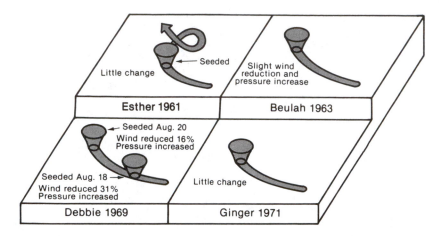

Figure 11-9. Results of seeding four different hurricanes are shown here. The most impressive results were obtained from Hurricane Debbie as winds were reduced and the central pressure increased following two different seedings.

Hurricane Beulah was seeded in 1963 with some indication of weakening winds. After a second seeding some decrease in wind speed was noted.

The third hurricane to be seeded was Hurricane Debbie in 1969. Seeding on August 18 was followed immediately by a 31% reduction in winds. It was not seeded on the 19th of August and the winds returned to their original intensity. Another seeding on the 20th resulted in the maximum winds dropping by 16%. These measurements have provided the best indication that seeding has some ameliorative effect.

A fourth hurricane, Ginger, was seeded in 1971. This storm had large oscillations of wind speeds and cloud conditions both before and after seeding. Seeding produced no changes greater than the natural variations. Since 1971, hurricanes have not been seeded through 1980. Project Stormfury will be continued with better equipped planes capable of obtaining continuous measurements. Hurricane modification would be largely beneficial since there is no indication that the beneficial rains produced by hurricanes can be eliminated by seeding operations.

Other suggestions for hurricane modification include the spreading of a monomolecular film over the ocean near the hurricane (Figure 11-10). This would interfere with the evaporation of water and trans-

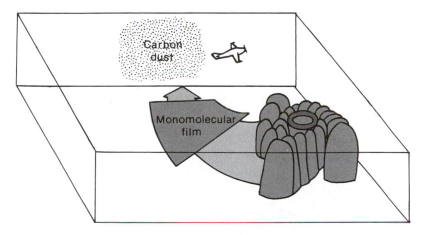

Figure 11-10. Other proposed methods of modifying hurricanes include spreading carbon dust into the atmosphere to absorb heat and the use of monomonecular film over the surface of the ocean. Neither of these methods have been used operationally.

port of moisture up into the storm, thus weakening the hurricane. A film of substance such as ethers and alcohols have the property of spreading in a very thin film over water to retard the loss of water to the atmosphere under ideal conditions. However, experiments have shown that wave action rapidly breaks up the film layer so this method of potential modification does not seem to be practical.

Another theoretical project is concerned with the addition of carbon dust to the atmosphere to change the temperature. It has been calculated that carbon dust spread in the atmosphere will absorb heat and create a heated layer of air. Thus the temperature structure of the atmosphere could be changed. It is estimated that 20 airplanes could spread sufficient carbon dust to cause a rise in temperature of $1°$ C/h for ten hours. A difference of $10°C$ could be enough to modify the structure of a hurricane. This method has not been tried because of several remaining questions including the environmental effects of carbon dust.

Hurricanes are very destructive storms with an average annual damage of \$440,000,000. A study conducted at the Stanford Research Institute addressed the problem of hurricane modification in a different way. It was assumed that hurricanes can be modified so the next question of the ramifications of this capability were considered. Both legal and economic repercussions are probable (Figure

11-11). A beneficial economic effect would result from reduced wind speeds. The amount of reduction in damage from hurricane winds as they are reduced can be estimated from statistics.

Wind speeds of about 100 km/h within a hurricane that strikes land correlate with average damages of $1,000,000. Hurricanes with winds of about 250 km/h are correlated with damages of $100,000,000. Since a relationship between property damage and wind speed exists, the economic value gained from seeding hurricanes can be estimated.

A probable legal repercussion of seeding hurricanes that strike land is numerous lawsuits from individuals experiencing damages from them. In spite of this factor it may be economically worthwhile to seed hurricanes because of greater benefits from reduced damages than from the combined loss from seeding operations and lawsuits.

HURRICANE SAFETY RULES

If hurricanes are a threat where you live, you should enter the hurricane season prepared. Each spring, recheck your supply of boards,

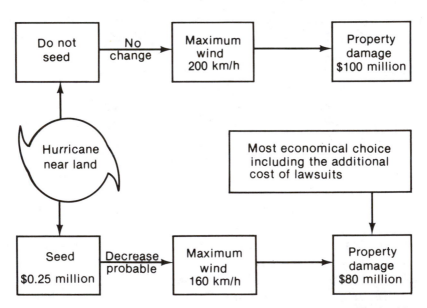

Figure 11-11. The decision whether or not to seed hurricanes that may strike land is not simple. However, if the maximum winds can be reduced by 20 percent, the corresponding damage to property can also be reduced. This could mean that the most logical choice would be to seed hurricanes even if the cost of seeding and possible lawsuits is included.

tools, batteries, nonperishable foods, and other equipment you will need if a hurricane strikes your town.

When your area is covered by a hurricane *watch,* continue normal activities, but stay tuned to radio or television for National Weather Service advisories. Ignore rumors. A hurricane watch means a hurricane may threaten an area within 24 hours. Continuously monitor the storm's position through Weather Service advisories. Check battery-powered equipment. A portable radio may become your only link with the outside world. Emergency cooking facilities and flashlights will be essential if utilities are interrupted.

Have your car fully fueled. If you own a boat, secure it before the storm arrives or move it to a safe area. When the boat is moored, leave it. Don't return to it once the wind and waves are up.

When your area receives a hurricane *warning,* additional activities should be completed. A hurricane warning means a hurricane is expected to strike an area within 24 hours. Continue to monitor the storms approach. It is time to board up windows or protect them with storm shutters. Secure outdoor objects that might be blown away or damaged, or bring them inside. Store drinking water; your town's water supply may be contaminated or diminished by hurricane floods.

Leave low-lying areas when advised to do so. If you live in a mobile home, leave it for more substantial shelter. Mobile homes are extremely vulnerable to high winds. If your home is sturdy and at a safe elevation, remain indoors during the hurricane. Because hurricanes often cause severe flooding as they move inland, stay away from the banks of rivers and streams. Tornadoes are often spawned by hurricanes and are among the storms' lethal effects. Therefore, when a hurricane approaches, listen to radio and TV for tornado warnings.

Hurricane watches and warnings are extremely important in reducing the loss of life from these storms. Since several hundred thousand people must be evacuated as major hurricanes approach land, specific evacuation routes are important. Such predetermined travel routes have been planned for many coastal communities with the help of the National Weather Service. Advanced preparation for hurricanes should also include a knowledge of the recommended routes to safety.

SUMMARY

The formation of a hurricane cannot be predicted but satellite photographs are used to identify hurricanes as they develop. The typical path of a hurricane in the northern hemisphere includes initial movement toward the west and northwest followed by a curving path that takes it northward and then toward the northeast. About 30% of hurricanes take more unusual paths that may include a loop.

The National Hurricane Center uses four different models to predict the future location of a hurricane. These are based on climatology, persistence, circulation, and dynamics. Satellite photographs are useful for obtaining information on the pressure and wind speed within a particular hurricane by comparing a satellite photograph of it with photographs of past hurricanes.

Project Stormfury is the national project including seeding hurricanes to modify their wind and pressure. The seeding of four hurricanes has given variable results, but provides some indication that the winds decrease and the pressure increases a few hours after a hurricane is seeded. Even though lawsuits may result from seeding a hurricane before it strikes land, it may be economically feasible because of reduced damages.

Hurricane watches and warnings are issued by the National Hurricane Center and relayed by the National Weather Service to the public. Appropriate preparations and reaction to hurricane warnings are important in saving lives from these storms.

12
Unusual Atmospheric Storms and Synoptic Patterns

As I watched the whirlwind travel across the barren field, I wondered about the strength of the winds inside and how long it would last. As I watched it make its way toward a farmhouse, I noticed that a dog was approaching the swirling air made visible by dust picked up from the dry ground. I watched in amazement as the dust devil surrounded the dog, lifted it into the air and took it up to a height of about 10 m before it reached the outer part of the vortex where it was lowered down to the ground, apparently without harm. As the dust devil moved closer to the mountains it appeared to intensify, and as it approached the highway it left little doubt that it had become a strong vortex as it turned a car around and set it in the ditch. The dust whirl that had started as a very small atmospheric disturbance was now so strong that human safety demanded that it be avoided. . .

DUST DEVILS

Dust devils are, in general, the smallest type of storm if they can be called storms at all. If you have ever traveled across the Desert Southwest in the summer, you have probably noticed dust devils or whirlwinds in action (Figure 12-1). They can develop from rising, unstable air by winds flowing around obstacles or by vortex twinning (Figure 12-2). Vortex twinning may occur as a vortex with a horizontal axis develops because of increasing wind speeds with height. As it hits obstacles, the vortex may be tilted to the vertical to form two dust devils where the ends touch the ground; thus the name vortex twinning.

Figure 12-1. A large dust devil in progress across the desert in Arizona is shown here. Such dust devils may vary greatly in strength. (Courtesy of Sherwood Idso.)

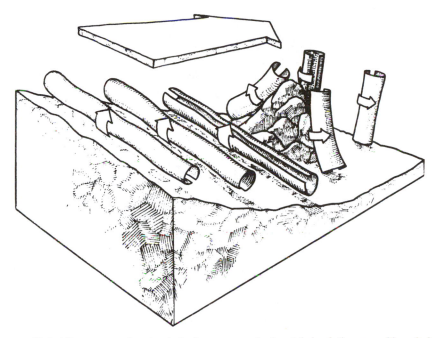

Figure 12-2. The process of vortex twinning occurs as horizontal circulation caused by wind shear is tilted into the vertical by topographic barriers. (After Sherwood Idso, *Weatherwise,* 1976.)

Warm surface air is important in dust devil generation just as in the generation of hurricanes. A further similarity exists in that both consist of spiraling air around a downdraft core. The dust devil, however, in contrast to the hurricane, revolves in either the anti-cyclonic or cyclonic direction. It is not generated by the larger circulation and is not affected by the Coriolis acceleration since it is so small. A major generating mechanism is surface heating, while the downdraft core simply replaces air that is rising in the surrounding vortex.

MOUNTAINADO

A slightly larger storm called a mountainado forms in some mountainous regions. Boulder, Colorado, has an unusually high frequency of winds greater than 100 km/h. Damage incurred cannot ordinarily be explained by straight winds. Houses along a particular street, for example, will have windows blown out, roof shingles lifted off with the nails pulled out, indicating suction similar to a tornado. Such damage appears to be associated with vertically oriented vortices containing winds of about 150 km/h.

Several such examples have been recorded in Boulder. On January 18, 1971, the wind damage pattern was such that it led to the conclusion that high velocity winds were accompanied by strong twisting action. On the 21st, according to an eyewitness account, the roof of a recently completed house was lifted 15 m into the air before it fell back to earth. The contractor who was working on a house next door described the wind as a circular motion dust devil.

A similar mountainado was described several years earlier by a man in Boulder who with his wife watched a strong whirlwind lift their 6 x 10 m shed about 10 m in the air to destroy it completely as it spiralled around in the air. Again, in 1975, on the 30th of November, during a time of relatively strong winds with 150 km/h gusts, vortices were observed to develop and move away from the mountains over snow covered ground. The mountainadoes were about 30 m in diameter and from the snow they picked up to make them visible, it was determined that they were about 150 meters in height. The horizontal winds in them were 75–150 km/h and usually averaged 75 km/h.

Since mountainadoes are generated in winter as well as summer, their formation mechanism is different from that of dust devils. The dynamics of air flow around an obstacle, such as the foothills, tend to initiate double vortices behind the barrier in the same way that the double vortex is initiated behind a strong updraft that blocks the airflow around a thunderstorm (Figure 12-3). Flow over the top of

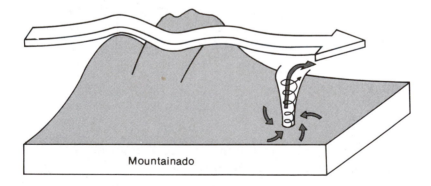

Mountainado

Figure 12-3. A vortex is intensified as air is extracted through its central core. Thus, a small vortex may be intensified until it has sufficient strength to damage houses and other properties. Mountainadoes are formed by airflow over topographic barriers interacting with a vortex that may extend to the surface.

the barrier intensifies the vortex just as strong upper air flow intensifies a tornado. Thus, we have the same mechanism on a small scale that generates a tornado on the large scale, i.e., horizontally circulating air with cooler air flow over it to intensify the updraft and strengthen the circular motion of the vortex.

THERMAL VORTICES

Atmospheric vortices are observed to form from thermal plumes from such sources as forest fires and industrial operations. An experiment set up in France by the Centre de Recherches Atmosphériques Henri Dessens uses an array of fuel oil burners, called the Météortron, to supply 1000 MW of heat. The resulting thermal plume regularly produces vortices. One method of vortex generation occurred as the rising plume interacted with light to moderate winds to form a double vortex structure (Figure 12-4). The two counterrotating vortices

Figure 12-4. Vortices may be formed by forest fires and other heat sources. Two vortices are formed here by the Meteorthron in France. Their formation mechanism is probably similar to the formation of a tornado by a double vortex thunderstorm. (Photograph courtesy of Christopher R. Church.)

may be formed in a similar manner to the formation of the double vortex thunderstorm on a large scale.

STORM METAMORPHOSIS

Some interesting transitions in the nature of storms occur that we might call storm metamorphosis. Each of the different individual atmospheric storms has its own unique structure depending on the formation mechanism. But in some cases, it is possible for one type of storm to transform into a different type and take on an entirely different structure (Figure 12-5). The hurricane is an example already noted since it may become a frontal cyclone. In order to do this, the structure has to be changed from a downdraft in the center of the hurricane to an updraft over a larger area. The size of the whole storm must increase along with a reduction in intensity. Frequently,

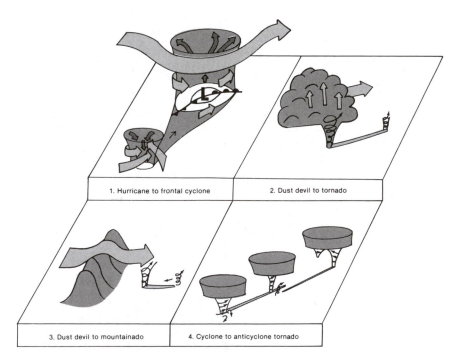

Figure 12-5. Storm metamorphosis occurs as hurricanes become frontal cyclones, dust devils become tornadoes or mountainadoes, and cyclonic tornadoes are converted to anticyclonic tornadoes.

such a metamorphosis produces a frontal cyclone that is much stronger than average as some of the characteristics of the hurricane are retained. Good examples of such storms were the frontal cyclones produced by the remnants of Hurricanes Carla, Agnes, and Camille.

Another example of storm metamorphosis is the dust devil transformed into a tornado. The dust devil is ordinarily very weak but it can connect with a fairly small developing cumulus cloud overhead to become a tornado. A developing cumulus cloud is composed primarily of a large updraft without the more organized structure of the large thunderstorm that typically produces tornadoes. However, the updraft can strengthen a vortex beneath the cloud into a tornado even though it could not generate a tornado without the preexistence of the vortex in its initial form as a dust devil. A tornado generated in this way is called an eddy tornado.

Dust devils may also become mountainados. In this case airflow over a topographic barrier interacts with a dust devil on the leeward side to transform it into a mountainado.

The transformation of a cyclonic tornado to an anticyclonic tornado is a less complete form of storm metamorphosis. Such transformations have been observed beneath a single thunderstorm. The mechanism for the transformation is probably similar to the laboratory observations that have shown that small anticyclonic vortices tend to form around a cyclonic vortex because of wind shear. As the cyclonic tornado decays, an anticyclonic vortex may connect with the updraft to become the tornado supported by the thunderstorm. Such transformations—and anticyclonic tornadoes in general—are rather uncommon.

UNUSUAL SYNOPTIC PATTERNS

Weather may also be unusual because of unusual synoptic air flow patterns. The meandering nature of the jet stream with warm air to the south and cold air to the north can produce record breaking high temperatures in one location, Anchorage, Alaska, for example, while another, such as Miami, Florida, is subjected to unusual cold waves (Figure 12-6). Since large meanders in the jet stream can persist for several days, and sometimes weeks, such unusual temperature anomalies can also persist.

Rapidly increasing air temperature is frequently related to warm air advection from the south. Occasionally record-breaking high temper-

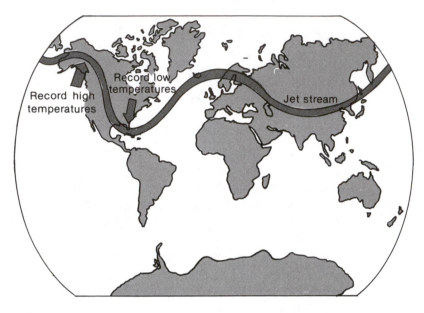

Figure 12-6. Unusual synoptic patterns such as large persistent meanders in the jet stream may cause record breaking high temperature in Alaska while temperatures may dip below freezing in Miami, Florida.

atures are produced by another mechanism—compression heating—as air descends within a high pressure region of the upper atmosphere. Such a synoptic pattern caused many long-term high temperature readings to be broken in North and South Dakota on April 21,1980. Figure 12-7 shows that maximum temperatures were quite high over much of North and South Dakota and parts of Nebraska and Minnesota. An air temperature greater than 37°C is extremely unusual this early in the spring especially so far north. It is apparent that the advection process is not the dominant factor in producing such high temperatures since the temperatures decrease southward across the Great Plains. The 500 mb map (Figure 12-7) shows the presence of a large ridge associated with a high pressure area over the Central United States that is conducive to subsiding airflow. This produces adiabatic heating and unusually warm temperatures if the air flow persists for a few days over a particular location as occurred in this example.

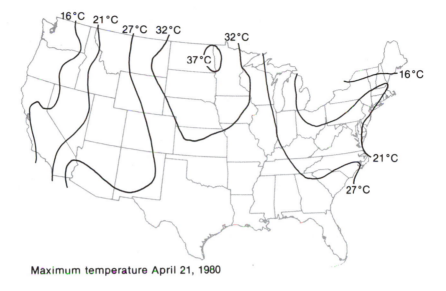

Maximum temperature April 21, 1980

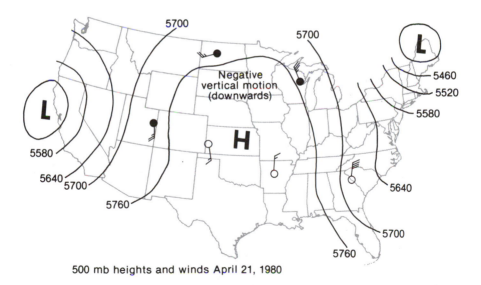

500 mb heights and winds April 21, 1980

Figure 12-7. Record breaking high temperatures associated with a heat wave of 1980 started early as the temperature soared to over 100°F in North and South Dakota in April. This heat was associated with a ridge in the jet stream, the presence of a high pressure area, and associated downward air currents that produced compression heating.

HEAT WAVES

Heat waves are forecast when the temperature is expected to rise above body temperature of 37°C. Continued temperatures above this level can cause death to humans, animals and vegetation. Heat waves can originate from advective or compression heating and are intensified by solar heating. The advection of warm air from the Desert Southwest may cause summer temperatures to exceed body temperature. Such advection is not usually as persistent as compression heating arising from subsiding air within a large high pressure system. High pressure systems are located southward from ridges in the jet stream and may stagnate for several weeks if the basic airflow patterns of the jet stream remain unchanged.

The weather of the whole Central United States was dominated by a persistent ridge in the jet stream and high pressure during most of June and July of 1980. The maximum temperatures persisted above body temperature for several consecutive days. Locations with such temperatures and the associated 500 mb heights are shown in Figure 12-8. By early July more than 1000 people had died from the heat and billions of dollars worth of summer crops were lost.

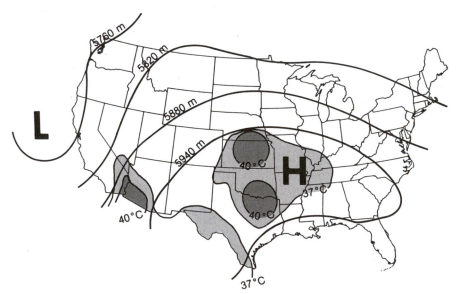

Figure 12-8. This synoptic chart for July 8, 1980, shows the typical heat wave pattern, including a ridge in the upper air flow and a surface high pressure system. This pattern persisted for much of the summer of 1980.

The hot weather is accompanied by lack of precipitation since the ridge in the jet stream and subsiding air associated with the high pressure generally prevent cloud formation; these contribute to more solar heating and continued hot weather. The parched land also contributes to hotter air temperatures as the cooling associated with evaporating water is absent. Thus, the temperature and precipitation elements become locked into an intensifying process that cannot be broken until the large scale airflow patterns change.

PERSISTENT SYNOPTIC PATTERNS

The weather may be very unusual because of persistent atmospheric synoptic conditions. The winters of 1977, 78, and 79 were much more severe than average as shown in Figure 12-9. Three consecutive winters of this severity have not been previously experienced since records have been kept in North America. The jet stream established itself in such a pattern that cold air from the north was continually fed into the United States as shown in Figure 12-10. Although some changes occurred within the jet stream flow pattern, the pattern rees-

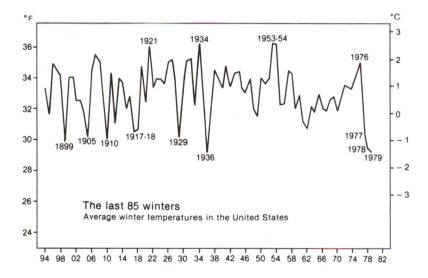

Figure 12-9. A plot of the average winter temperatures in the United States over the last 85 years shows the severity of the three consecutive winters 1977–1979. Other single years such as 1936 and 1899 were about as cold but were followed by much warmer winters. (After *Weatherwise*, Feb. 1980.)

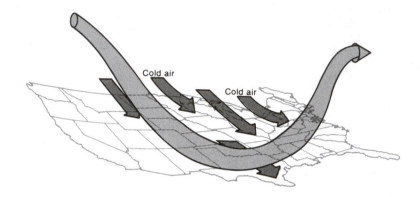

Figure 12-10. Cold air outbreaks are typically associated with a trough in the jet stream. The north to south upper-air current helps pull in additional cold air from the north, and surface pressure systems originated within the trough of the jet stream are usually followed by north winds behind the center of lowest pressure.

tablished itself to continually feed cold air into the United States. A large number of long wave cyclones also formed and moved across the United States during these winters. They brought blizzard conditions and more snowfall even though the air temperature was already quite cold.

UNUSUAL INTENSITY

Some atmospheric storms are unusual because of their intensity. Hurricane Camille was an example of a hurricane in this category. Some of the strongest winds and highest ocean levels ever recorded in the United States were attained during the landfall of Hurricane Camille. Hurricane Camille was a small hurricane, but certainly compensated for her size by her intensity.

Some tornadoes are unusual because of their large size and very strong winds. Tornadoes in this category were the Topeka, Lubbock, Omaha, and Wichita Falls tornadoes. These were all about 1 km in diameter and contained winds so strong that houses were completely demolished.

Some frontal cyclones are so strong that they can be classified as unusual; one of these occurred in February 1978. The central pressure in this frontal cyclone dropped to 958 mb. The storm began to become more symmetrical and approached the appearance of a hurricane with some indication of the development of an eye in the center. This storm dumped over 0.5 m of snow on New York City. The snow was continuous for more than 40 hours; businesses were closed and cars were abandoned on the Long Island Expressway. Over one 15 km stretch more than a thousand cars were abandoned.

During this same month, February 19, 1978, an extreme high pressure system was also observed; the pressure reached 1050 mb. Although such anticyclones do not have unfavorable weather associated with them, except for greater pollution potential, they are quite unusual when the pressure increases to this magnitude.

Unusual precipitation intensities have occurred that are worth noting. In Holt, Missouri, in 1942, over 30 cm (12 in) of rainfall fell in about one hour's time. Hailstones have accumulated to depths of 18 inches from single thunderstorms with drifts more than 1 m deep. Snow depths of 2 m from single storms have been recorded. Chicago received more than 1 m of snow from one storm during the severe winter of 1978. Various other North American weather extremes are shown in Figure 12-11 for pressure, temperature, rainfall, wind, snow, fog, and hail.

UNUSUAL TYPES

The weather we have may also be unusual because of its type. An example is the supercell thunderstorm that produces tornadoes and hail because of its organized double vortex internal structure. Most thunderstorms are smaller, begin with an updraft then switch to a downdraft and dissipate. This makes the tornado producing thunderstorm unusual.

A large nontornadic thunderstorm can create winds of hurricane speeds that rip roofs off houses and drive rain so hard that travel on highways is impossible during the onslaught. Such winds may come from two different sources within a thunderstorm. One source of strong winds is the downburst of rain-cooled air through the region of anticyclonic rotation within the leading northeastern part of the thunderstorm. This downdraft is strengthened as it connects with the major updraft area to the southwest of the downdraft region (Figure 12-12). Surface locations between the two vortices within

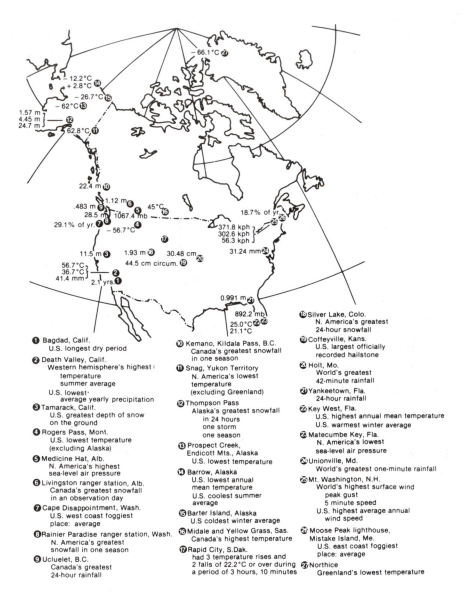

❶ Bagdad, Calif.
U.S. longest dry period

❷ Death Valley, Calif.
Western hemisphere's highest:
temperature
summer average
U.S. lowest·
average yearly precipitation

❸ Tamarack, Calif.
U.S. greatest depth of snow
on the ground

❹ Rogers Pass, Mont.
U.S. lowest temperature
(excluding Alaska)

❺ Medicine Hat, Alb.
N. America's highest
sea-level air pressure

❻ Livingston ranger station, Alb.
Canada's greatest snowfall
in an observation day

❼ Cape Disappointment, Wash.
U.S. west coast foggiest
place: average

❽ Rainier Paradise ranger station, Wash.
N. America's greatest
snowfall in one season

❾ Ucluelet, B.C.
Canada's greatest
24-hour rainfall

❿ Kemano, Kildala Pass, B.C.
Canada's greatest snowfall
in one season

⓫ Snag, Yukon Territory
N. America's lowest
temperature
(excluding Greenland)

⓬ Thompson Pass
Alaska's greatest snowfall
in 24 hours
one storm
one season

⓭ Prospect Creek,
Endicott Mts., Alaska
U.S. lowest temperature

⓮ Barrow, Alaska
U.S. lowest annual
mean temperature
U.S. coolest summer
average

⓯ Barter Island, Alaska
U.S coldest winter average

⓰ Midale and Yellow Grass, Sas.
Canada's highest temperature

⓱ Rapid City, S.Dak.
had 3 temperature rises and
2 falls of 22.2°C or over during
a period of 3 hours, 10 minutes

⓲ Silver Lake, Colo.
N. America's greatest
24-hour snowfall

⓳ Coffeyville, Kans.
U.S. largest officially
recorded hailstone

⓴ Holt, Mo.
World's greatest
42-minute rainfall

㉑ Yankeetown, Fla.
24-hour rainfall

㉒ Key West, Fla.
U.S. highest annual mean temperature
U.S. warmest winter average

㉓ Matecumbe Key, Fla.
N. America's lowest
sea-level air pressure

㉔ Unionville, Md.
World's greatest one-minute rainfall

㉕ Mt. Washington, N.H.
World's highest surface wind
peak gust
5 minute speed
U.S. highest average annual
wind speed

㉖ Moose Peak lighthouse,
Mistake Island, Me.
U.S. east coast foggiest
place: average

㉗ Northice
Greenland's lowest temperature

Figure 12-11. Various weather extremes in North America. (After US Army Engineer Topographic Laboratories, Fort Belvoir, Virginia.)

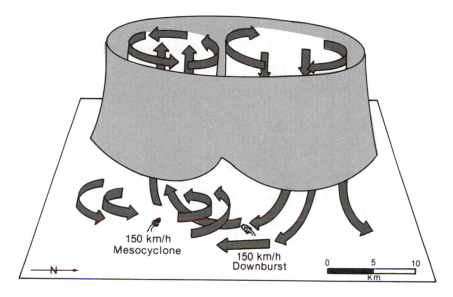

Figure 12-12. Strong damaging winds may be produced in different areas of a single thunderstorm. Within the leading northern part of the thunderstorm, a downburst of strong winds may be sufficient to produce damage. In the southern part of a severe thunderstorm the cyclonic rotation may extend to the ground even though a tornado is absent. These winds may also be strong enough to produce damage.

the thunderstorm can experience winds greater than 150 km/h as this circulation is completed. These winds are typically from the north or northeast.

Another region of high wind speeds is located beneath the cyclonic rotation of the mesocyclone within the thunderstorm (Figure 12-12). These circular winds have a diameter of 5 to 10 km; thus they are often mistaken for straight winds of 150 km/h or more. Because of their association with the cyclonic rotation within the thunderstorm, these winds form in the southern part of a thunderstorm.

The tornado producing cyclone is also unusual because of type. It is a cross between the longwave and trough cyclones. It is stronger and produces intense weather because of its location just north of the jet stream where it can take advantage of the higher wind speeds in the upper atmosphere as they flow over the fronts that extend southward (Figure 12-13).

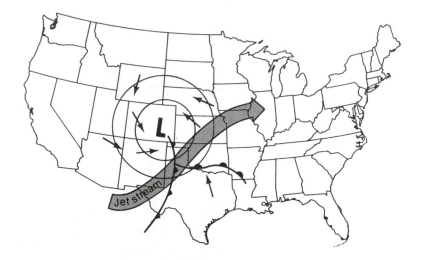

Figure 12-13. The frontal cyclone and jet stream location are shown here for the weather system that produced the Wichita Falls tornado in 1979.

UNUSUAL COMBINATIONS

Other weather systems are unusual because of combinations of various factors. A good example is the Arizona haboob (Figure 12-14). This is an Arabic word from the Sudan meaning "to blow." Desert regions are affected by this particular storm which is unusual because it consists of a very dense wall of dust which may reduce visibility to less than 1 m. The sudden onset of a haboob was one of the factors that contributed to the failure of the American effort to rescue hostages from Iran in 1980.

Phoenix, Arizona, often experiences these storms. They are caused by winds greater than about 100 km/h that whip up loose dust from the desert. After the wind is laden with dust it causes damage by sandblasting. This removes paint, and kills trees and people. On July 16, 1971, a thunderstorm that caused a haboob moved out of the Santa Cruz river valley southeast of Tucson toward Phoenix at about 50 km/h. The temperature dropped by 10°C and the relative humidity rose from 33% to 74%. Nine people were killed and 27 injured on highways because of reduced visibility. Phoenix has an average of 2 or 3 of these storms every year with as many as 12 in a single year.

The haboob is caused by the gust front of a thunderstorm which combines with the dust of the desert to produce the haboob. The

Figure 12-14. The haboob is an intense dust storm produced in desert regions by the gust front associated with a strong thunderstorm. Storms such as this one observed near Phoenix, Arizona, can reduce the visibility to near zero and cause many types of problems. (Courtesy of Sherwood Idso.)

gust front forms from the downdraft within the thunderstorm and spreads out ahead of it. Wind speeds of 100 km/h are common. If rain occurs with the thunderstorm it is behind the haboob.

MISCELLANEOUS UNUSUAL WEATHER

In southern California a warm dry wind similar to the chinook wind off the eastern slope of the Rockies blows over the coastal mountain ranges (Figure 12-15). When the pressure gradient favors east winds the flow of air through the mountain passes sometimes becomes concentrated and reaches speeds of 150 km/h. The descending air is warmed adiabatically and may cause a rapid increase in temperature along with a rise in fire hazard since the relative humidity of the air also decreases rapidly as the air heats. These winds are called Santa Ana winds after a particular city near Los Angeles where they are unusually strong.

Other unusual types of weather are interesting, such as colored rainfall. The red rains of France and blue rains of Greece are caused by a combination of dust particles which serve as condensation

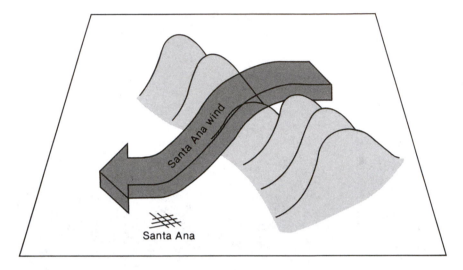

Figure 12-15. Strong winds are occasionally experienced at Santa Ana, California, as the pressure distribution forces air through particular mountain passes. As this air descends it is heated by compression and becomes a warm dry wind called the Santa Ana wind.

nuclei within a thunderstorm. If the dust particles are colored, the rain droplets may appear colored. In some desert areas dust particles are red because of iron content. Copper minerals cause blue or green rain. Yellowish-red rains in the Central United States fall from clouds that have picked up red dust from Oklahoma or Texas.

The weather may be unusual in a variety of other ways. On March 26, 1895, a thunderstorm produced a tornado that moved across Albany, New York, with accompanying snow instead of rain. Two years earlier, on December 16th, a tornado moved through Oswego, New York, accompanied by sleet or ice pellets. It is apparently never too cold to snow. At Fort Yellowstone, Wyoming, records indicate that 7 cm of snow fell on the 2nd of February 1899 when the maximum temperature did not get above $-18°$F all day. In contrast it snowed on the 4th of July in 1879 in Portland, Maine.

Thunderstorms that produce snow with associated thunder and lightning are unusual, but it is even more unusual for them to also produce St. Elmo's Fire. However, this was observed at Fort Huron, Michigan, on March 25th, 1930. During a fall of heavy wet snow, St. Elmo's Fire was observed at the end of the arrow of the wind vane and on the axis of the vane and the tail. It made hissing sounds loud enough to be heard 20 m below the wind vane.

SUMMARY

Various types of vortices form in the atmosphere. Some of these may be considered to be unusual because they are not as well known or understood as other more common atmospheric vortices. Dust devils are surface generated vortices that form on hot summer days particularly in desert regions. In mountainous terrain, the mountainado is a vortex that occasionally becomes strong enough to damage houses and other property. The mountainado forms as the air flowing over a topographic barrier interacts with a vortex to strengthen it.

Thermal vortices associated with forest fires and with oil burners designed especially to generate them are of considerable interest. It is not uncommon for a double vortex structure to be generated within thermal plumes. Such a structure may be analogous to the generation of the double vortex structure within thunderstorms.

Many atmospheric storms undergo metamorphosis. When this happens one of the individual atmospheric storms with its own unique structure is changed into a different type of atmospheric storm. Thus, hurricanes change into frontal cyclones, and dust devils may change into mountainadoes or tornadoes under the right conditions.

Unusual synoptic patterns may be responsible for very different weather in various parts of the world. Large meanders in the jet stream may provide Alaska with warmer than normal temperatures at the same time that Florida is experiencing below freezing temperatures. Heat waves are associated with a particular type of upper air flow pattern. A persistent ridge in the jet stream is common with heat waves. This allows a surface high pressure system to develop and persist for an extended period of time. Within a high pressure system the subsiding air undergoes compression heating as it descends giving rise to hot air near the surface.

The weather may also be unusual because of persistent synoptic patterns, unusual intensity, or types of storms. The intensity of some hurricanes and tornadoes is such that they must be considered unusual because of this. For example, Hurricane Camille was one of the strongest hurricanes ever to strike the United States, similarly the large tornadoes such as those that struck Lubbock, Topeka, and Wichita Falls were unusual because of their intensity.

Various weather elements and surface features may combine to create unusual weather occurrences. Thus, the haboob is caused by the gust front of a thunderstorm which picks up dust from desert areas and generates a dust storm with very low visibility. Many other types of unusual weather are possible. The red rains of France and the blue rains of Greece are caused by dust particles which serve as condensation nuclei within thunderstorms. The rain then takes on the color of the particles that serve as nuclei for the raindrops.

13
Floods and Drought

As the highway entered the canyon we noticed that it followed the small stream ahead. The stream looked slightly swollen but this was of little concern. Suddenly, as darkness closed in, our headlights met a wall of oncoming rushing water. The next thing we knew, we were struggling to get out of the van as it was carried along by the racing current. The headlights became useless as they sank beneath the muddy water; we were now fighting for our lives within the raging current. In almost total darkness, with sheer canyon walls on both sides, the two of us tried desperately to stay together clutching at any debris that might help keep us afloat.

As the current carried us against one of the walls of the canyon, we clawed at the smooth rock desperately searching for a way out of the torrent. As my hand touched a ledge, I grabbed Jim and we desperately struggled to drag ourselves out of the water. We found that the ledge was only large enough for one person and even then, it required effort to stay on it. We soon discovered a ledge of similar size just above us, and Jim proceeded to climb to it.

After more than an hour on the ledge, I felt something hit my arm and lodge there. As I picked it up I realized it was the radio transmitter we normally carried. Knowing that it could prove to be a very important piece of equipment in our circumstances, I made sure that I held on to it during the rest of the sleepless night. After an eternity, the morning sunlight began to light the canyon and we realized we were more than 10 m above the ground on the only ledges in the whole canyon wall within sight. As we finally contacted civilization through the transmitter we realized that no one was going to believe we had survived such a flood on these tiny perches. . . .

FLOODS

Few areas of the world are free from at least an occasional flood. Even the Desert Southwestern United States occasionally has a thunderstorm that produces very heavy localized runoff. Figure 13-1 shows a comparison of various areas of the United States troubled by floods. The North Atlantic region is most prone to flooding and major flooding occurs throughout the Missouri River Basin and other large basins.

Severe floods in the United States result from a number of different types of weather systems. Hurricane Agnes, in June of 1972, was responsible for the greatest economic flood loss in the history of the United States. Over 4 billion dollars worth of damages were caused by Hurricane Agnes and 105 lives were lost. The second greatest eco-

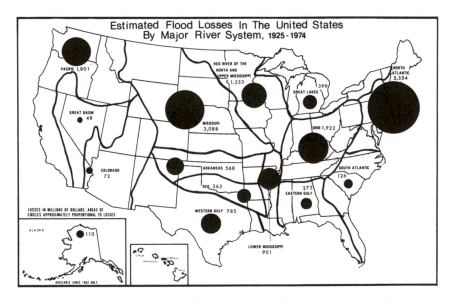

Figure 13-1. The flood losses in major river systems based on several-year records show considerable variation in distribution. The North Atlantic region, the Missouri River Basin, and the Pacific Coast all experience considerable flood losses. (Climatological Data, National Summary.)

nomic flood loss developed the following year, when the entire Mississippi River system was flooded. Over one billion dollars worth of damages were reported and 33 lives were lost. The third most damaging flood was in July of 1951 when the Kansas and Missouri Rivers flooded. These caused over 900 million dollars worth of damages and 28 lives were lost (Figure 13-2).

The greatest killer flood in the United States was in the Ohio River Valley, in March of 1930, when 467 people were killed. The Mississippi River Valley flooding in the Spring of 1927 resulted in 313 deaths. The flood in Rapid City, South Dakota, in June of 1972 and the Big Thompson, Colorado, flood in July 1976 both resulted in more than 200 deaths.

Figure 13-2. This photograph taken on July 14, 1951 shows the severity of flooding in Kansas City. The foreground includes a residential section with the roofs of some of the houses still visible above the water. (Official US Navy Photograph.)

Causes of Floods

The primary cause of floods is extraordinarily heavy precipitation, but a smaller amount of precipitation may also produce flooding when it falls on ground that is already saturated. Precipitation from any given storm is partitioned into the amounts disposed of by infiltration into the soil, surface detention, evaporation, and runoff. The amount of runoff into streams is, of course, the major factor in producing floods. The infiltration rate is related to soil characteristics. If the soil has a coarser texture, water soaks into it more rapidly than if it has a finer texture. Management practices such as cutting vegetation or overgrazing also affect the rate at which water will be absorbed into the soil.

Infiltration rates may vary from 5 cm per hour at the beginning of a storm to less than 1 cm per hour as rainfall continues. Stream flow is composed of a certain amount of base flow that represents the ground water flow into a stream. Surface runoff during a rainstorm adds to this base flow. Most streams or rivers have a definite floodplain (Figure 13-3). This is the low lying ground surrounding the borders of the river. The river may flood this lowland once per year, or once per decade, depending on the river channel and amount of precipita-

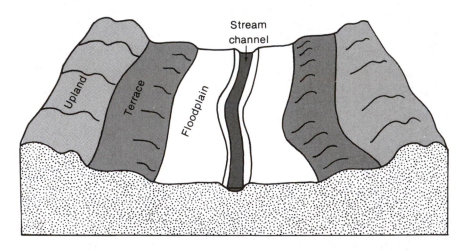

Figure 13-3. Rivers flow through a stream channel and are surrounded by a relatively flat area forming the floodplain. Outside the floodplain, terrace and upland regions exist. Flooding is normally confined to the floodplain with occasional flooding of terrace locations.

tion. Above floodplains are broader steplike expanses of terrain called terraces. These may flood each decade or century.

The greatest observed precipitation rates are of interest. Over 3 cm (1.23 in) of rain fell in one minute in Unionville, Maryland, and more than 30 cm (12 in) of rain fell in only 42 minutes in Holt, Missouri. These extreme rainfall rates are much more than needed to produce substantial flooding.

The greatest economic flood loss in United States history was produced by the remnants of Hurricane Agnes in 1972, as previously stated. During the week preceding Hurricane Agnes, frontal activity brought soaking rain to the mid-Atlantic region, from New England to Virginia. Showers and thunderstorms dumped as much as 12 cm of rainfall on this area. In Central Pennsylvania 6 cm were common. Throughout much of New England, the soils were already on the way to saturation before Agnes arrived. Florida also had rainfall for two or three days prior to the passage of Hurricane Agnes. Floods developed along the Gulf as Hurricane Agnes formed in the Gulf and approached land. Among the highest total rainfall amounts were about 20 cm at Maple and Tallahassee, Florida. The flooding as Hurricane Agnes struck land was much less than after it was modified into a frontal cyclone. As it moved northward, and was modified when it interacted with the jet stream, as previously described, it began to dump heavy rain into Virginia and much of New England. This produced severe flooding on the James and Appomattox River basins and along the Potomac and smaller rivers.

In the eastern half of the James River basin and western half of the Appomattox basin, flooding was the worst in history. Crest stages exceeded those of Hurricane Camille in 1969 and topped high water records dating back two hundred years. The average rainfall over the whole James River basin was 15 cm from the 19th to the 22nd of August 1972. New crest stage records were set at many cities along the James River. The river swamped a 200 block area of downtown Richmond in the worst flood in the city's history. The crest level set at the city locks on the 23rd of August topped the old mark set back in 1771.

The Potomac River began flooding on the 22nd of August. In downtown Washington, D.C., at Wisconsin Avenue, a flood crest of more than 4 m lasted for about 8 hours. In Pennsylvania, Agnes' heavy rains also fell on wet ground. Small streams began flooding first, followed by flooding of the major rivers. The Susquehanna River exceeded previous flood stage marks set in March of 1936 (Figure 13-4). Crests were generally more than 4 m above flood stage. At Wilkes-Barre, the water came over the levees completely flooding the city with a crest more than 6 m above flood stage. Thus, Hurricane Agnes produced major flooding, not as a hurricane, but in the form of an intense frontal cyclone.

Repeated Frontal Cyclones

A very different meteorological condition was responsible for the second greatest economic flood loss in United States history. The flooding of the Mississippi River in 1973 was caused by repeated frontal cyclone passage through the Central United States. The jet stream flow patterns are very influential in determining the amount

Figure 13-4. Hurricane Agnes produced this flood in Harrisburg, Pennsylvania, on June 27, 1972. The water rose more than 3 m above flood stage from the overflowing Susquehanna River. (Official US Coast Guard Photograph.)

of precipitation over large areas. Repeated frontal cyclone passage of both the longwave and trough type occurred over the Mississippi River basin from the 20th of March 1973 through the 9th of April 1973 (Figure 13-5). This brought widespread rain over the Central United States. By the end of March, the Mississippi River was above flood level from Clinton, Iowa, to Donaldsonville, Louisiana, with some additional flooding at stations upstream as far as Minnesota. Major flooding also occurred along the lower Missouri, Ohio, and other tributary streams in the Mississippi Valley.

During April the severe overflow of the Mississippi River continued (Figure 13-6). This flooding was aggravated by additional periods of heavy rainfall from frontal cyclones during April. The Mississippi River rose to record high levels from Burlington, Iowa, to Cape Girardeau, Missouri. Millions of acres of rich farmland were inundated throughout the Mississippi Valley. Tens of thousands of people were evacuated from their homes. Entire populations of some communities were evacuated; damage to roads, buildings, and bridges was extensive.

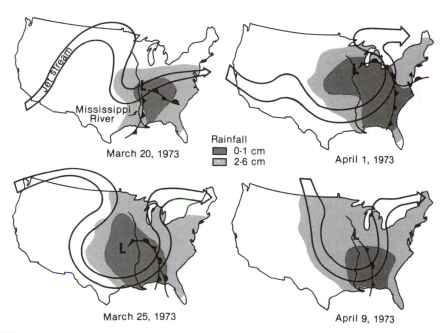

Figure 13-5. The Mississippi River flood of 1973 was produced as frontal cyclones repeatedly soaked the river basin during March and April.

Stationary Front

Another flood producing meteorological condition is established as a well defined front stalls and becomes stationary. This happens when either a warm or cold front becomes parallel to the axis of the jet stream. Thus, the upper level winds are along the front rather than

Figure 13-6. Many cities along the Mississippi River were flooded in 1973. This photograph taken April 30, 1973, includes the main street of West Alton, Missouri. This town located about 10 km from the confluence of the Missouri and Mississippi Rivers became the joining point when a number of levees broke on both rivers. The picture was taken four days before the rivers crested. There was considerable concern at this time that the rivers may have cut a new channel through parts of the town. (Official US Coast Guard Photograph.)

across it to move it along. If the stationary front has persistent south winds southward from it and north winds northward, the mechanism is established for producing precipitation in the same locality for an extended period of time. These were the meteorological conditions that produced the flood responsible for the third largest economic flood loss—the Kansas City flood in 1951.

During the 72 hour period, July 9 to 12th, excessive rainfall fell over much of the eastern half of Kansas (Figure 13-7). Much of the Kansas River Basin received more than 25 cm of water on top of previously wet soil from the 25 cm or more of rainfall during June that fell over all of northeastern Kansas. This caused severe flooding of cities along the Kansas River, including Kansas City, Missouri and Lawrence, Kansas.

Individual Thunderstorms

It is not too unusual for a single thunderstorm to produce flashfloods that result in loss of lives. Thunderstorms with a capacity to cause flood conditions can occur throughout almost any of the tropical or midlatitude locations. Even the Sahara Desert is subjected to an occasional thunderstorm. Some of the greatest killer floods have resulted from the rainfall of thunderstorms.

The synoptic patterns for developing flood producing thunderstorms in mountainous areas are quite different from those that produce thunderstorms with heavy rain in the Great Plains. If the jet stream is absent and the upper level winds are light, weak southeasterly surface winds help set the stage for heavy rainfall in mountainous areas (Figure 13-8). Such a synoptic pattern developed over Colorado on July 31, 1976. A flash flood was produced in the Big Thompson River Canyon 40 miles north of Denver, Colorado. The Big Thompson River ordinarily is less than 1 1/2 m deep but at 6 pm on July 31st, it started to rain and within the next six hours thunderstorms had produced 35 cm of rainfall. By 9 pm a wall of water about 10 m deep was rushing down the Big Thompson Canyon taking everything with it. It caused 50 million dollars worth of damage and killed more than 200 people. The opening paragraphs of this chapter were based on the accounts of two survivors of this flood.

The surface and upper atmospheric synoptic charts for July 31, 1976, show a surface front located over Colorado. The 500 mb winds

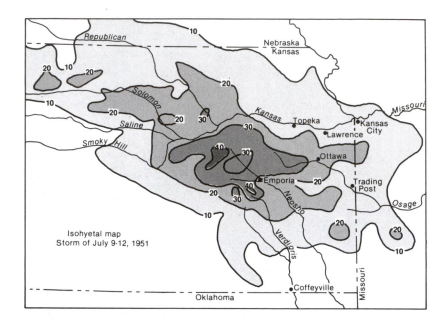

Figure 13-7. The top map shows the number of centimeters of water that fell over eastern Kansas on July 9–12, 1951. Major flooding occurred in all low-lying areas. The bottom photograph shows north Lawrence on July 14, 1951. (Official US Navy Photograph.)

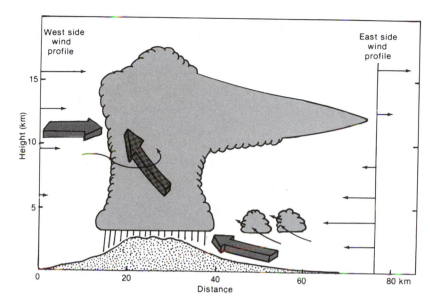

Figure 13-8. Flash floods in mountainous terrain are caused by very different synoptic conditions than occur over the plains. A thunderstorm may produce heavy rainfall in a localized area when the upper level winds are very light and opposite to the low level air currents.

were very light with a ridge in the jet stream flow patterns and a general high pressure area located over the Great Plains area (Figure 13-9). Thus thunderstorms building over the Big Thompson Canyon were not moved along by the upper airflow and were fed by the low level winds to give a continued downpour for several hours. Although the rain was very localized, and the heavy rain extended over a distance less than 20 km, it was centered over the mountainous terrain that fed the Big Thompson River.

On June 9, 1972, a similarly devastating flood occurred in Rapid City, South Dakota. The surface and 500 mb synoptic charts were amazingly similar to those causing the Big Thompson flood (Figure 13-10). A cold front pushed through South Dakota on June 9th beneath a ridge in the upper airflow pattern with very light winds. In 1972, South Dakota's operational cloud seeding program was just getting into full swing. A rain gage network had been set up at strategic locations to measure the effect of seeding clouds in the Rapid City area. More than 250 kg of salt were used for seeding clouds near Rapid City between 3 and 5 pm on June 9th. Heavy rainfall began about 5 pm and thunderstorms produced more than 20 cm of water

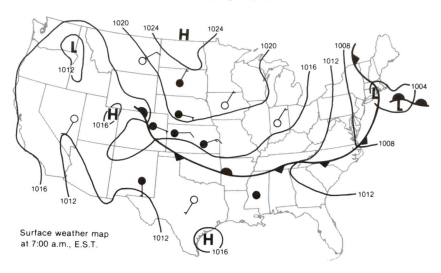

Sunday, August 1, 1976

Surface weather map
at 7:00 a.m., E.S.T.

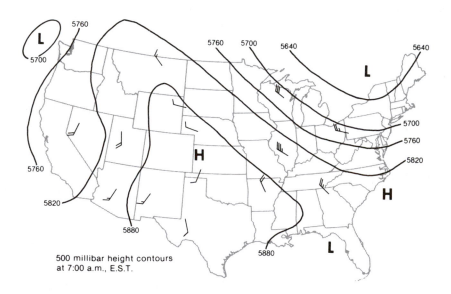

500 millibar height contours
at 7:00 a.m., E.S.T.

Figure 13-9. The surface and upper air maps corresponding to the Big Thompson flood in Colorado show very light upper level winds from the west and surface winds from an easterly direction. Such synoptic conditions are typical of those producing flash floods in mountainous areas.

Friday, June 9, 1972

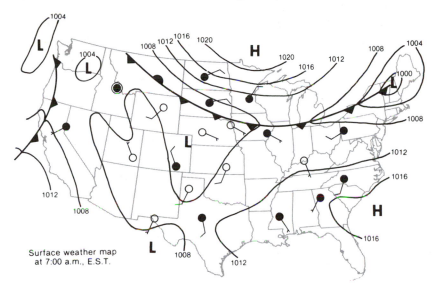

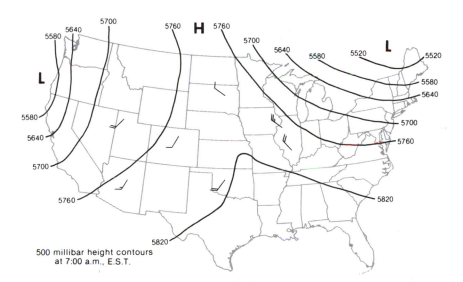

Figure 13-10. The surface and upper air maps for the Rapid City, South Dakota flood show light upper level winds, light surface winds and the presence of a front. These conditions are typical of those producing flooding in mountainous terrain.

in some localities near Rapid City (Figure 13-11). All the seeded area received more than 5 cm of rainfall.

Although all the natural factors for producing flood conditions were present in South Dakota on June 9, 1972, the fact that weather modification was attempted within the area resulted in loss of public

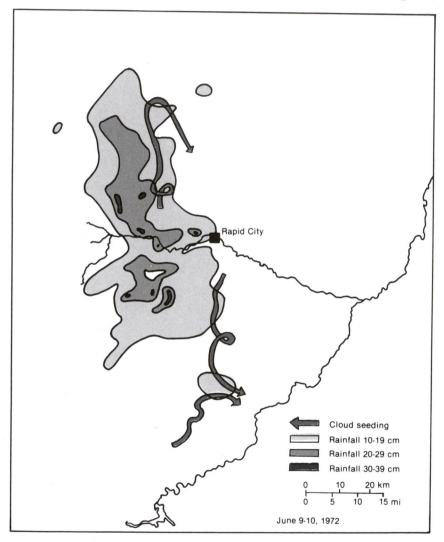

Figure 13-11. Cloud seeding was conducted only a few hours prior to the Rapid City flood. Even though all the factors were present to produce the flood from the natural weather factors, this seeding operation had a detrimental effect on weather modification in South Dakota (After Armand, *Report on Rapid City Flood of 9 June 1972.*)

support for the operational weather modification program in South Dakota.

Snowmelt and Other Causes

Various other weather factors that may be involved in producing flash floods include a slow moving frontal cyclone. If a weather system moves very slowly then it affects the same area for a longer period of time; thus, the potential for flood conditions is increased.

Rapid increases in temperature may be responsible for causing floods during the spring if a heavy snow cover is on the ground. Rainfall on top of a snow cover is also a frequent flood producing condition. Frozen ground may also be a factor during the spring months. If the warming in the spring occurs rather rapidly, and the ground is still frozen underneath, the water does not infiltrate into the ground and therefore a greater amount of it goes into the runoff segment of the hydrologic cycle. Thus, various combinations of snow, temperature, and rainfall may be responsible for flood conditions.

Flood Warnings

The National Weather Service operates a flash flood warning service. Flash flood watches and warnings are prepared by the Weather Service Forecasting Office (Figure 13-12). They utilize such information as the quantitative precipitation forecast prepared by the National Meteorological Center. Quantitative precipitation forecasts give the expected amount of precipitation within the next 24 hrs. These are used along with such other information as satellite photographs, radar, and surface observation networks. River stations may also monitor critical water levels for several kilometers upstream from cities (Figure 13-13). When a flash flood watch or warning is issued by the National Weather Service, it is transmitted to local newsmedia for appropriate community action.

DROUGHT

The importance of water is rapidly emphasized if it becomes unavailable. Drought is a deficiency of water and can be best defined in terms of departure from normal precipitation. A county or city becomes accustomed to the amount of rainfall it normally receives,

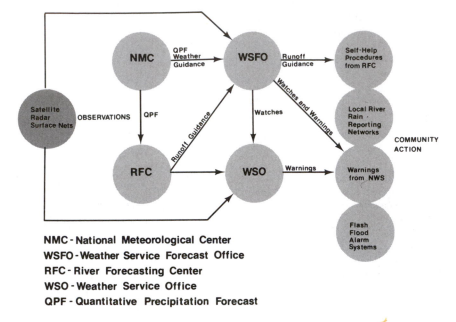

NMC - National Meteorological Center
WSFO - Weather Service Forecast Office
RFC - River Forecasting Center
WSO - Weather Service Office
QPF - Quantitative Precipitation Forecast

Figure 13-12. Flash flood watches and warnings are prepared by the Weather Service Forecast offices to be relayed to the public for appropriate community action. (After *Operations of the National Weather Service.*)

and all water related activities are conducted accordingly. When departures from normal amounts occur, they may create severe problems. The 1930s are well known for dust storms (Figure 13-14). More recently southern California experienced below normal amounts of rainfall during the winter months of 1976 and 77. The drought became so severe that by early 1977 water was rationed because of the shortage.

A drought in the early 1960s in the northeastern United States was the worst in 160 years. It brought widespread unemployment and critical water shortages throughout most of the heavily populated Eastern United States. In New York City, restaurants no longer offered a glass of water. The use of water by individuals was also restricted; people were not allowed to water lawns or wash cars, even the water hydrants were locked tight to prevent access to this water source. By 1965, the reservoirs of the city of New York were down to 25% of their capacity and the reservoirs of Washington, D.C. forced residents there to face the real possibility of water rationing.

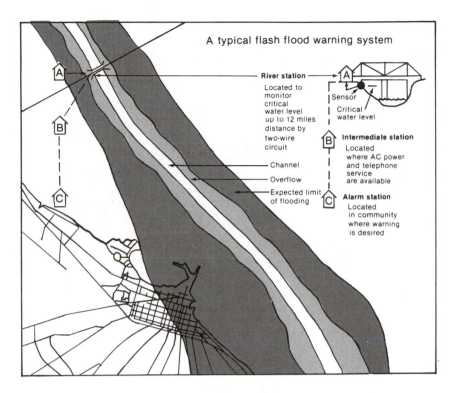

Figure 13-13. Flash flood warning systems may include sensors to measure the water level at critical locations and detailed information on the channel of the river and height of flood stage. (After *Operations of the National Weather Service.*)

Many different types of losses result from drought. Obvious losses are agricultural crops, trees, and grass (Figure 13-15), but many additional kinds of losses are equally important. These include business losses such as lost income from investments, lost revenues to cities because of lack of water, and many other miscellaneous losses. It is, in fact, difficult to determine the exact economic significance of drought because of its nature. For example, the onset of drought in a distant country may affect the price of bread on the other side of the world.

Drought tends to be exaggerated as civilization expands into deserts as far as possible. Continual drought exists in deserts, but we call this condition aridity. However, on the fringes of deserts the water supply is marginal and is therefore adequate in some years and absent

Figure 13-14. The dust within a dust storm such as this one of the 1930s became so dense that it was a threat to life and property. (Courtesy of the Library of Congress.)

or deficient in others. Thus, normal variations in rainfall have very disastrous results in semiarid regions.

Drought Producing Weather Patterns

Very dry and hot weather conditions are generated in the Central and Eastern United States by a ridge in the jet stream (Figure 13-16). Such a flow pattern was present during the summer of 1965 as extreme dry conditions were experienced in the Eastern United States. A ridge also became stationary over the Central United States in 1976 and 1980 to produce extremely dry weather. If the jet stream is displaced farther northward, this leaves very light upper level winds over the United States with resulting hotter and dryer weather.

Along the West Coast, drought is frequently influenced by the position of the Hawaiian high pressure area. This semipermanent high pressure region normally shifts southward during the winter to allow frontal activity to bring rain to the West Coast. If, on the other hand, this high pressure area remains further northward during the winter months the result is dryer than normal weather. Since much

Figure 13-15. Extreme soil erosion can result from a combination of dry soil, lack of vegetation, and strong winds. A South Dakota wheat farm is shown above as observed in August 1970. (Courtesy of USDA Soil Conservation Service.)

of the West Coast receives almost all of its rainfall during the winter months, the position of the Hawaiian High is crucial for their water supply.

In tropical regions the intertropical convergence zone is important in determining which areas will receive rainfall. As the intertropical convergence zone moves with the seasons, the converging air between the northeasterly tradewinds of the northern hemisphere and the southeasterly tradewinds of the southern hemisphere cause deep convective rain clouds. The movement of the intertropical convergence zone varies from year to year bringing variability of rainfall amounts. A primary factor of the Sahelian drought of the early 1970s was the displacement of the intertropical convergence zone about 2° latitude further southward than normal across Africa.

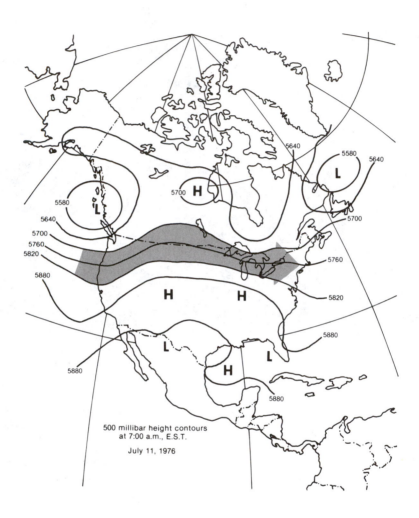

Figure 13-16. Drought is typically produced from a ridge in the upper airflow pattern, such as that shown above for July 11, 1976. Surface high pressure systems may be persistent beneath an upper level ridge and also contribute to continued dry weather.

Agricultural Drought

The effect of dry weather on agriculture is determined not only by the amount of rainfall but by the amount of water lost to the atmosphere. The maximum water loss rate is called the potential evapotranspiration. This is defined as the amount of water lost from actively growing crops with an abundant supply of moisture in the soil. The potential evapotranspiration rate is determined primarily by meteorological variables such as temperature, relative humidity, winds, and radiation, with the greatest influence coming from the temperature and humidity.

The actual evapotranspiration at any time is governed by still another factor—the amount of soil moisture. Since abundant water is often not available for plant use, the actual evapotranspiration rate is frequently less than the potential rate. The distributions of actual and potential evapotranspiration are shown in Figure 13-17. The potential evapotranspiration is more than 200 cm per year in much of the desert southwestern United States. Across the northern United States, it is only about 50 cm. The actual evapotranspiration rate is much less than the potential rate in those areas where precipitation is small, thus the Desert Southwest loses less than 25 cm per year, while the Gulf Coast states lose more than 100 cm per year.

Water balance diagrams can be made by plotting the monthly values of precipitation, and potential and actual evapotranspiration. Such diagrams are useful means of describing the overall moisture conditions for a region. Specific water balance diagrams for several locations are shown in Figure 13-18. If the potential evapotranspiration rate exceeds the amount of water available from precipitation and soil moisture a deficiency of moisture exists. The magnitude and duration of moisture deficits are of major importance to agriculture. If soil moisture deficits exist for an extended period of time, they may have a devastating effect on agricultural production.

The southwestern United States has the greatest accumulated deficiency of moisture. In addition to the amount of rainfall, the variability is extremely important in semiarid regions. The variability of rainfall in the United States as shown in Figure 13-19 is greater in dryer areas than in wetter areas. The amount of precipitation also varies with time. Some indication exists for a 20 year cycle in rainfall variability. The 1930s were extremely dry in the Central United States;

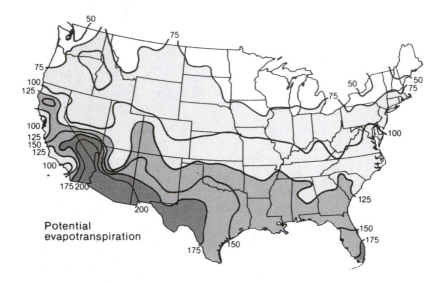

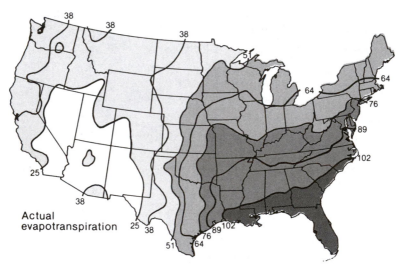

Figure 13-17. The potential and actual evapotranspiration (cm) shows considerable variation throughout the United States.

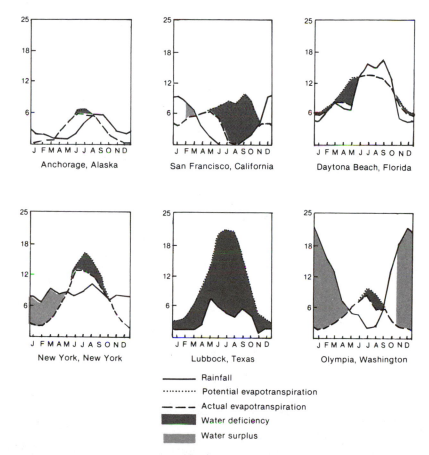

Figure 13-18. Water balance diagrams constructed by plotting the monthly rainfall, potential evapotranspiration, and actual evapotranspiration rates (shown above in cm) are useful in showing the moisture characteristics of a region. (After Eagleman, *The Visualization of Climate*, D.C. Heath and Co., Lexington, Mass., 1976.)

drought conditions prevailed in the 1950s with some dry years in the 1970s.

Continued drought can change the landscape (Figure 13-20). The dust storms of the 1930s covered fences with mounds of dust; a dust storm 500 km in diameter can carry 100 million tons of dust. Dust storms are caused by a combination of lack of precipitation and little or no soil cover. Dust storms are not limited to the 1930s; a recent dust storm occurred on February 23, 1977. As a strong longwave cyclone traveled eastward it picked up dust from Colorado, New

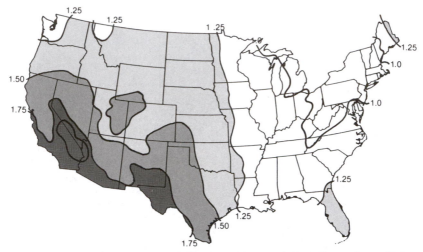

Figure 13-19. The variability of rainfall expressed by a statistic similar to the coefficient of variation shows that rainfall is most variable per inch in the Southwestern United States. It is least variable per inch of rainfall in the Northeastern United States (After Eagleman, *The Visualization of Climate,* D.C. Heath and Co., Lexington, Mass., 1976.)

Figure 13-20. The extreme dust storms of the 1930s completely changed the landscape in some locations. (Courtesy of the Library of Congress.)

Mexico, and Texas and carried it along south of the longwave cyclone center across the southern United States and into the Atlantic. The dust concentration was large enough to reduce the visibility to almost darkness in the middle of the day. The dust also showed up on satellite photographs (Figure 13-21).

Recent dust storms have not matched those of the 1930s, however. From 1935 to 1937, unusually strong winds combined with the extended drought to produce great clouds of blowing top soil which obscured the sun over much of New Mexico, Colorado, Oklahoma, Kansas, and Texas. The grim experiences of the 1930s revealed that the semiarid lands of the Great Plains could not be farmed by the same methods as the farmlands in the more humid East (Figure 13-22). Many farming practices have changed accordingly. New machinery is used that leaves the soil surface less disturbed, terraced farming is practiced, and more crop residues are left on the ground to help stabilize the soil and help it to absorb and retain more moisture.

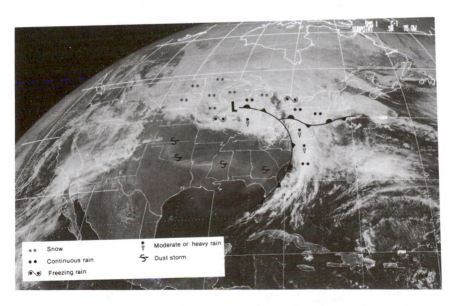

Figure 13-21. A recent dust storm is shown in this satellite photograph taken February 24, 1977. The light area extending through Oklahoma, Southern Arkansas, and other southern states is caused by extremely dense, blowing dust.

Figure 13-22. The devastating effects of the dust storms of the 1930s have not been repeated in part because of adjustments in farming practices in semiarid regions. (Courtesy of US Department of Agriculture.)

In the early 1950s, from 1950 through 1956, drought again returned to Texas, Oklahoma, New Mexico, Arizona, Kansas, Missouri, Colorado, and Nevada. An analysis of tree rings, at the University of Arizona, brought the conclusion that the drought of the 1950s was the worst to afflict the Desert Southwest in 700 years. More than half the farmers in much of the afflicted area had to find jobs off the farm. The national disaster of the 1930s was not repeated, however, because the national economy was strong and growing with plenty of jobs in towns and factories.

SUMMARY

A nation's water supply is probably its most important asset; however, it does not arrive from the sky in an evenly distributed and measured quantity. Floods are caused by several different specific meteorological factors. Repeated frontal cyclones were responsible for the flooding of the Mississippi River in 1973. The ground became saturated and as the rains continued more water ran into the streams and rivers creating severe flood conditions along the Mississippi River.

If a front becomes stationary, it can also cause flooding. This was responsible for a very damaging flood in 1951 in Kansas City. Flash floods may result from individual thunderstorms. A single thunderstorm that drops several centimeters of water within a few hours may create flood conditions. This occurred in the Big Thompson Canyon in July of 1976, and also in Rapid City, South Dakota, in 1972.

Several other factors including snow melt may cause or contribute to the generation of floods. Flash flood watches and warnings are prepared by the National Weather Service. Since flash floods are caused by thunderstorms that are localized events their forecasting is somewhat comparable to forecasting tornadoes. It is important to know that flash floods can occur suddenly and are, in general, smaller in scale than the distance between reporting weather stations.

At the other extreme in moisture supply, drought conditions are a major problem in many parts of the world. Drought conditions are generally produced by specific upper air flow patterns in the atmosphere. A ridge in the jetstream developed over the Eastern United States in the 1960s and was responsible for drought conditions there. A ridge in the jetstream that persists for several weeks decreases the chance for rainfall.

The effect of drought may be disastrous on agriculture, since vegetation requires plenty of water for maximum yield. If the evapotranspiration rate requires more water than is available from precipitation and moisture stored in the soil, crops suffer and yields are reduced. Severe crop losses and dust storms formed in the 1930s. These were generated by a combination of lack of rainfall, poor farming practices, and strong winds. Such dust storms still occur but are, in general, less frequent and less intense.

Part 3
Weather Simulation, Modification, Management and Human Response

14
Urban and Agricultural
Weather Modification

As I approached the airport the director's voice came in loud and clear, "Runway three is closing in again. Give me another 20 g of silver iodide." Since this was the runway I would need for landing the small plane, I was happy to comply. As the silver iodide flare burned, the thin layer of fog began to clear behind the plane. My thoughts turned to the large thunderstorm developing to the south, and I wondered if it would respond so obediently to the treatment we would soon be giving it. . .

URBAN WEATHER MODIFICATION

The management of weather in urban areas is to a large extent inadvertent except for fog dispersal over airports and a few other extreme examples such as the Botanical Garden in St. Louis, Missouri, where a dome is used to control the microclimate as required for tropical vegetation. A larger and more famous example is the Astrodome in Houston where an entire stadium is enclosed to remove all possibility of a sports event being rained out.

Other more subtle, inadvertent modifications of the urban climate are very common. These include changes in the amount of solar radiation, the temperature in urban areas in comparison with rural settings, the amount of moisture in the air, and wind direction and speed. Some investigations indicate the precipitation is also affected by urban activities. In addition to these changes in primary climatic elements, the bioclimate is also altered.   Thus the composite effect of the climate on people and vegetation in urban areas is different from rural areas because of human activities.

Radiation

The amount of radiation at the surface of even small cities is decreased. The measured distribution of radiation in downtown Lawrence, Kansas (population 50,000), showed 15% less solar radiation in comparison to rural areas (Figure 14-1). An even larger area has a 10% reduction in solar radiation. One of the reasons for this is the presence of foreign material in the atmosphere; particles and gases that allow less solar radiation to reach the surface. Measurements in larger cities show that the reduction of radiation is even greater. In Kansas City, with over 1 million people, the measured radiation was 20% less than in rural areas. Radiation measurements were obtained from an instrument mounted on the roof of an automobile in such a way that it always measured vertical radiation while the vehicle was in motion. Transects were followed in two directions through the city to obtain the geographical distribution of solar radiation and other variables.

Temperature

Since the radiation is decreased in cities, you might expect the temperature within a city to be lower than in rural regions. However, this is not the case. An urban heat island or increase in temperature in downtown areas commonly exists (Figure 14-2). The urban heat island develops from a variety of causes; partly because of changes in the radiation budget, not only solar but also infrared radiation. In addition to this, urban areas are constructed from materials with very different physical properties from the soil and vegetation that was replaced. These construction materials do not have a high moisture content, whereas soil retains water and provides cooling as the water evaporates from it. Concrete, brick, roofs, and asphalt surfaces in urban areas absorb heat during the day and hold it during the night.

Measurable differences in temperature exist even around small buildings. Measurements around a shopping center showed that within it the temperature was 1°C warmer than in surrounding areas only a very short distance away.

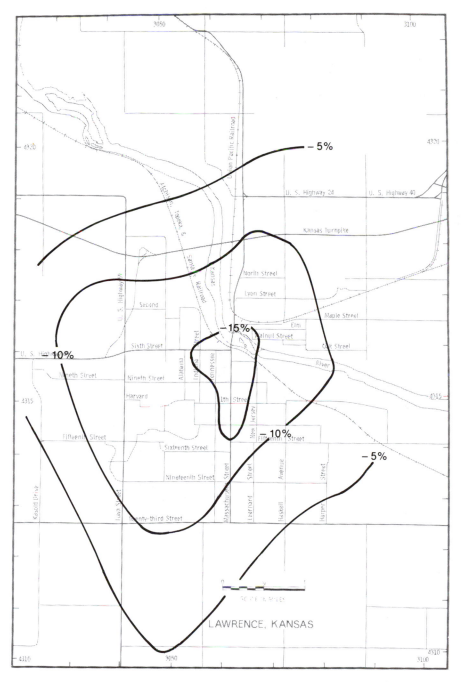

Figure 14-1. Solar radiation reduction in downtown Lawrence, Kansas, amounted to 15% with smaller reductions in surrounding areas.

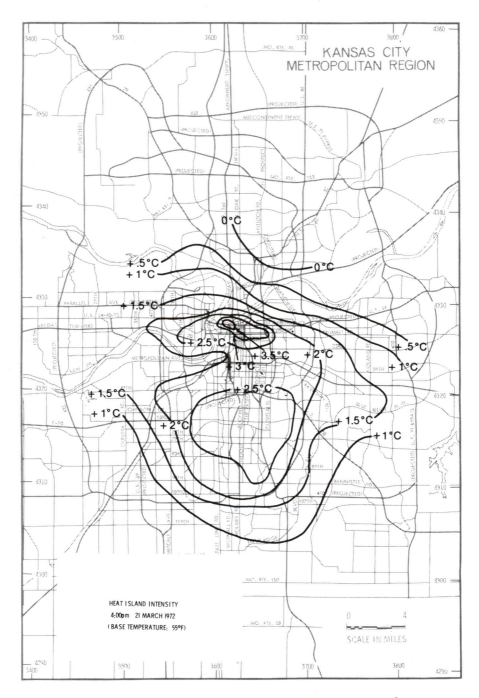

Figure 14-2. The urban heat island in Kansas City was measured to be 3.5°C at 4 pm on March 21, 1972.

Measurements over various seasons of the year in several cities show that the heat island exists in all seasons, both day and night.

Morning and afternoon temperature measurements are shown in Figure 14-3 for Lawrence, Kansas. The largest difference in temperature between downtown and rural areas was 4.5°C while the smallest was 0.5°C, and the average was about 2°C.

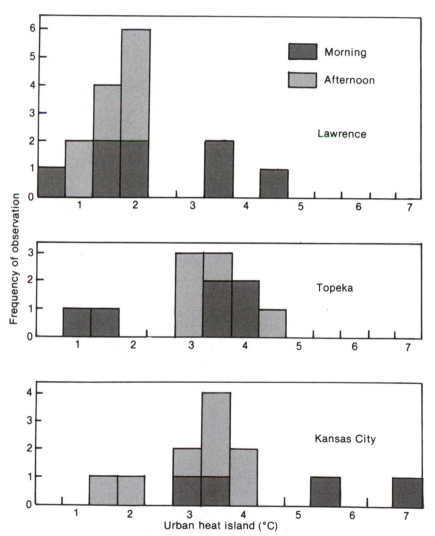

Figure 14-3. The urban heat island is greater in larger cities, such as Kansas City, than smaller ones. It is present in both morning and afternoon hours. (After Eagleman, *Atmospheric Environment*, 1974.)

As the size of a city increases, modification of the urban climate also increases. In Topeka, the maximum measured heat island was 4.5°C, the minimum was 1°C and the average was 3°C. In a still larger urban area, Kansas City, the intensity of the heat island ranged from 7°C down to 1.5°C with an average heat island of 4°C. Thus, a relationship between the size of the city and the magnitude of the urban heat island exists.

The urban heat island produces several side effects. One of these is a reduction in the cost of heating houses during the winter for all areas within the urban heat island. Since a straight line relationship exists between the amount of gas consumed for heating and the outside air temperature, each degree increase in outside air temperature reduces the amount of required heating (Figure 14-4). By the same token, greater amounts of energy are required to cool buildings on hot days during the summer months.

Relative Humidity

The relative humidity of city air is much less than in rural areas (Figure 14-5). In Lawrence, the average reduction in the relative humidity is about 5%. Since the relative humidity is determined by the air temperature in addition to the amount of water vapor in the air, the reduction in relative humidity is primarily caused by the increase in temperature. Measured decreases in relative humidity averaged 9% in Topeka (Figure 14-6), with a maximum of 14%, and a minimum of 4%. The reduction in Kansas City varied from 22% to 6% with an average of 13%. Since greater urban heat islands occur in larger cities, the relative humidity reduction is also most pronounced in larger cities. The magnitude of the temperature and relative humidity differences between cities and rural areas are measurable both in the morning and evening.

Human And Vegetation Environment

One of the effects of the altered urban climate is expressed in human comfort; downtown areas are more uncomfortable on a hot sunny day. This effect can be computed by using the comfort index based on an empirical equation using dewpoint temperature (°F) and air temperature. The comfort index = 0.4 X (temperature + dewpoint temper-

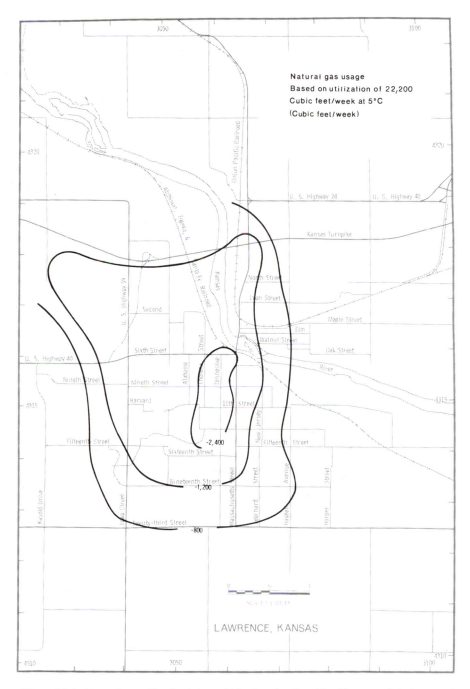

Figure 14-4. Natural gas utilization is less within the urban heat island because of the increased temperature.

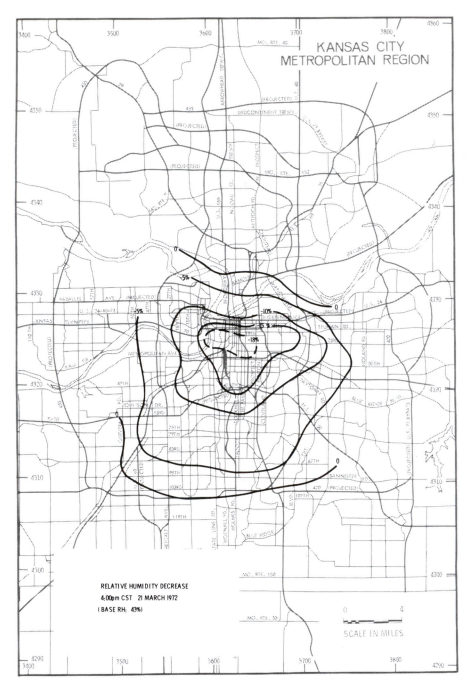

Figure 14-5. The relative humidity is less in downtown areas primarily because of the increased air temperature.

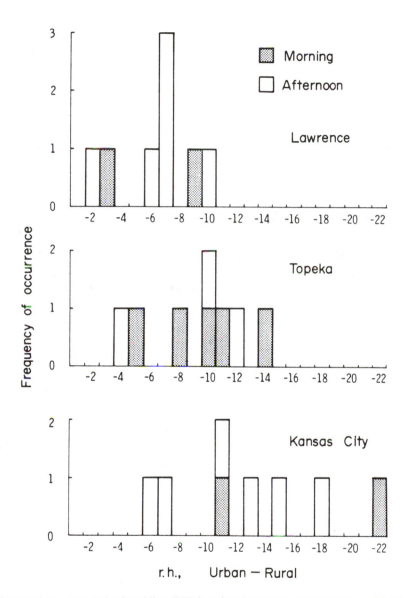

Figure 14-6. The relative humidity (RH) is reduced a greater amount in large cities than in smaller ones. (After Eagleman, *Atmospheric Environment*, 1974.)

ature) + 15. The higher the number, the greater the discomfort. With a value above 80, most people are uncomfortable; at 75, about 50% of the people are uncomfortable; between 70 and 75, less than 50% experience discomfort.

The hotter more uncomfortable environment downtown contributes to health problems in some cases. Heat waves such as those of the summer of 1980 may cause critical health problems especially for older people. On the other hand, the central business district provides a more comfortable environment on cold, calm days due to the heat island.

The urban environment for vegetation is also changed from that in surrounding rural areas. Since the temperature is higher and humidity is lower the evapotranspiration rate from vegetation is greater. Thus, the water requirement for vegetation is greater in downtown areas. The growing season is also a few weeks longer in urban areas; flowers bloom earlier in the spring and killing frosts come later.

These various inadvertent modifications to urban weather are all greater in larger cities and less in smaller cities. The degree of modification can be quantified by use of satellite photographs to measure the urbanized area. When this value is compared with weather measurements taken from urbanized areas of different size, it is apparent (Figure 14-7) that a good relationship exists. For larger urban areas, the magnitude of the heat island increases in a relatively straight line. The rate of evapotranspiration and the comfort index also increase; only the relative humidity decreases, relative to rural areas, as a city grows.

Wind

Data are scarce on the effects of an urban area on wind characteristics. An obvious effect is that tall buildings are obstacles to airflow which tend to reduce the wind speed. They may also channel the air through avenues that will then suffer from increased wind speeds. Some measurements show that breezes tend to flow in toward the center of the city (Figure 14-8). This could be expected from the effects of the urban heat island. If it is hotter over the center of the city, the air will rise and bring air toward the city from surrounding rural areas.

Composition of the Atmosphere

The composition of the atmosphere differs between urban and rural areas.   More particles and gaseous pollutants are present in the urban atmosphere, and greater problems arise from a combination of stagnant weather systems and pollutants in cities.   Classic pollution episodes developed in Donora, Pennsylvania, in October of 1948, and in Lon-

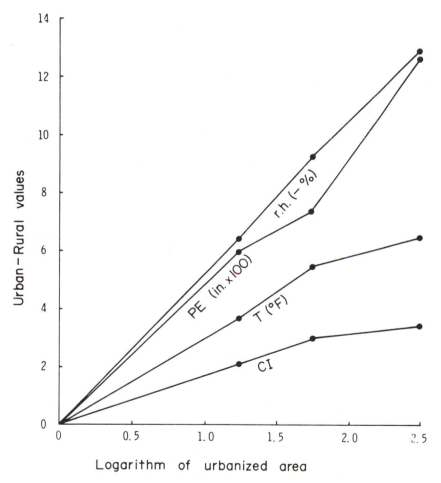

Figure 14-7. The size of the urbanized area is related to the urban climatic modifications. The lines above show the difference in urban and rural values for the four climatic variables: relative humidity (RH), potential evapotranspiration (PE), temperature (T), and comfort index (CI). The urban minus rural difference increases for all four of the variables except relative humidity, which decreases as the size of the urbanized area increases. (After Eagleman, *Atmospheric Environment,* 1974.)

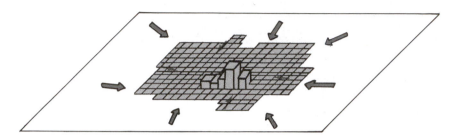

Figure 14-8. Under light wind conditions breezes tend to blow toward the city because of the heat island effect.

don, England, on December 6, 1952, when thousands of people became ill and scores of people died. In both cases the air patterns were a factor with very light winds south of a ridge in the jetstream. Upper air inversions were also associated with these episodes. Inversions cause a cessation of the normal vertical mixing of the atmosphere and trap pollutants near the ground. Los Angeles is continually plagued by pollution problems because of a persistent high pressure system with an upper air inversion. This limits vertical atmospheric mixing while the mountains to the east limit horizontal mixing. Thus particulates and gases such as nitrogen dioxide, carbon monoxide, and sulfur dioxide increase in concentration. Some of these such as nitrogen dioxide react with water vapor and other gases in the presence of sunlight to form additional compounds including some that are eye irritants.

Precipitation

Precipitation is much more variable over short distances than any other weather element. Therefore it is difficult to evaluate the effect of a city on the amount or type of precipitation. A few investigations indicate that the amount of rainfall is greater over some cities. At La Porte, Indiana, 5 year averages of precipitation show a good correlation with the amount of haze at Chicago. The pollution in Chicago apparently affected the weather downwind from it. Cities located near lakes, rivers, or changing topography are not ideal candidates for evaluating such a relationship, however. Most investigations of the rainfall near cities where most of these other influencing factors are eliminated reveal insignificant changes in the amount of precipitation, although visibility is frequently less and fog is more frequent.

Comparison of the rainfall in Kansas City with that of other cities of varying size in Kansas and Missouri revealed that the amounts of rainfall were similar but the intensity of rainfall was less in Kansas City (Figure 14-9). More frequent lighter rains fell in Kansas City than in surrounding regions. More abundant condensation and deposition nuclei would account for this difference.

One investigator found some indication of hail distribution modification around St. Louis since slightly more hail fell downwind from St. Louis than in the city itself (Figure 14-10). The number of summer hail days in St. Louis from 1949–1965 was 8, while in Edwardsville, Illinois, it rose to 15 days; an increase of almost two times the amount in St. Louis.

Summary of Inadvertent Urban Modifications

It is clear that for some weather variables, urban areas have a definite influence; for others, it is much more difficult to determine the specific

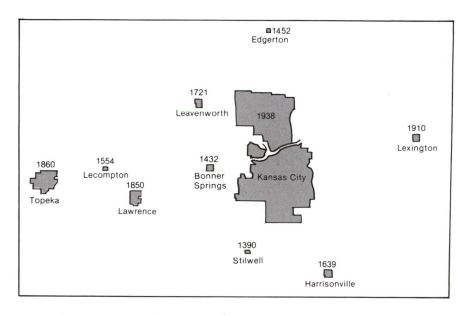

Figure 14-9. Although rainfall amounts were not altered noticeably within the Kansas City region, the number of days with precipitation in the 20 year period 1949–1968 shows that the larger cities, Kansas City and Topeka, experienced more rainy days. Thus, the intensity of rainfall was altered.

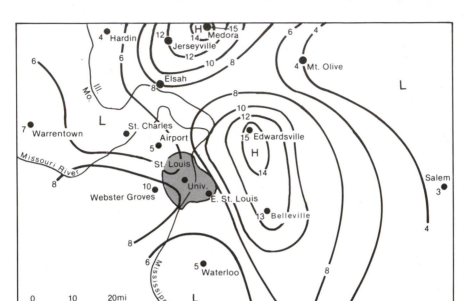

Figure 14-10. The number of hail days during the summer increased downwind from St. Louis according to statistics based on the period 1949 to 1965. (After Changnon, 2nd Conference on Urban Environment, AMS, 1972.)

nature of the relationship. Precipitation may be increased by more abundant particulate matter; on the other hand, too many particles may decrease the amount of rainfall. Solar radiation and relative humidity are decreased in urban areas while the temperature, evapo-transpiration, comfort index and length of growing season are all increased.

Many city planners are modifying these alterations of city weather by using more trees, parks, and green belts to break up the effect of urbanization on the city climate.

Intentional Urban Weather Modification

While many of the urban modifications are inadvertent and come about because of urbanization, a few weather modification activities are conducted through deliberate design. Since fog over airports and bridges can delay travel, modification programs are carried out at many locations (Figure 14-11). In general, those geographical locations

Figure 14-11. Fog dispersal programs are carried out at some airports and, on an experimental basis, over bridges, such as that shown here. Warm fog was cleared from over the bridge across Smith Mountain Lake, Virginia, by a helicopter. (Courtesy of V.G. Plank, *Weatherwise,* June 1969.)

with cold fogs at temperatures of −5°C to −20°C can be seeded to dissipate the fog and rapidly improve visibility. This is important for both large and small aircraft. Fog seeding programs are carried out on a regular basis at more than 40 airports. The cost of seeding fog is small compared with fog related losses, which amounted in 1971 to $15,000,000 with an additional cost to passengers of $28,000,000 for example. Over the United States, the cost is estimated at $75,000,000 lost each year by airlines due to fog problems.

AGRICULTURAL WEATHER MODIFICATION

Appropriate air temperature and sufficient rainfall are critical for crop production and world food supply. A poor wheat yield means higher prices for bread and meat. Orange groves and peach orchards are very sensitive to cold temperatures as the fruit starts to develop. The amount of water is important to vegetation at all times, but is more critical at some stages; as corn kernels begin to form, for example.

Modification of Water Supply

Conventional means of responding to the water requirement of crops consist of such responses as selecting particular crops that are suited to the normal climate of the region, or adding water by irrigation systems. A more recent and more dramatic technique for adding to the water supply is cloud seeding. Cloud seeding is accomplished by using either silver iodide or dry ice with the idea of converting super-cooled water to ice crystals that have a reduced vapor pressure and can grow in an environment where water droplets would evaporate (Figure 14-12). The basis of cloud seeding is the experiments conducted in 1948 by Vincent Schaefer, who discovered that a very small number of dry ice crystals would create millions of additional ice crystals in a cold chamber with supercooled liquid droplets. This process is called homogeneous nucleation since the dry ice is so cold that ice crystals form spontaneously. It was soon discovered that

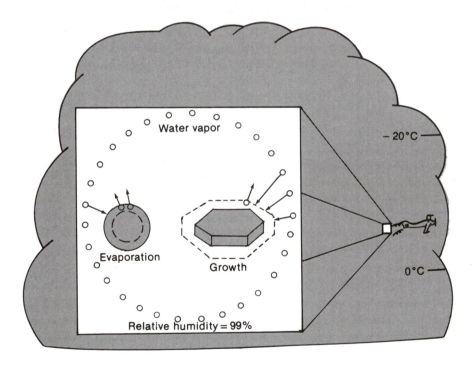

Figure 14-12. Cold cloud seeding involves converting supercooled water to ice crystals that can grow in the same environment where liquid water droplets would evaporate.

other materials can be used if introduced at the proper atmospheric temperature. The most used is silver iodide, since it has a structure very similar to the six-sided ice crystal. Since no cooling occurs with silver iodide the temperature of the cloud must be from $-5°C$ to $-25°C$. Within this range much of the cloud is composed of liquid water droplets that can be converted to ice crystals.

Cloud seeding with dry ice is accomplished by spreading dry ice ground into fine particles through the cloud. Silver iodide is burned and the smoke particles contain the deposition nuclei which are introduced into the cloud. One gram of silver iodide will create billions of deposition nuclei. The amount of seeding material that is introduced may determine the outcome and the optimum range for either dry ice or silver iodide is not clearly known. The most effective application method is directly into the cloud from an airplane, but silver iodide burners are sometimes set on the ground with the hope that air currents will carry the seeding agent into the cloud at an appropriate rate.

Warm clouds are seeded with a different seeding agent; ordinary finely ground salt. Salt is an effective condensation nuclei. However, the atmosphere usually has an abundance of condensation nuclei even though it may have a deficiency of deposition nuclei for the formation of ice crystals.

The evaluation of cloud seeding operations is not a simple procedure (Figure 14-13). Either warm or cold cloud seeding results could be checked by seeding one cloud and not seeding another identical cloud;

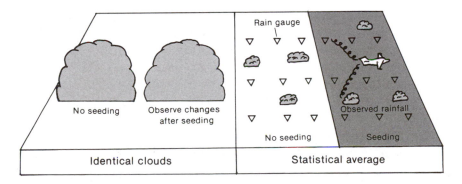

Figure 14-13. The results of cloud seeding could be checked by seeding one of two identical clouds, but since these are impossible to find statistical averages are normally used to evaluate cloud seeding experiments.

but identical clouds are either rare or absent. An alternative is to seed an area that includes several clouds for comparison with an unseeded area. Rain gages located at the surface then measure the differences. This common approach requires a large sample to overcome the uncertainties accompanying such experiments. It has been calculated that an increase in rainfall of 10% would require 67 cases in order to gain statistically valid correlations. If the increase in rainfall is less than 10% even more tests would be required for verifications. The results of cloud seeding experiments are somewhat mixed, but good indications exist for assuming an increase in precipitation of about 10% if cold clouds are seeded.

Some concern has been expressed over the environmental ramifications of applying silver iodide to clouds. Investigations have been conducted to check for accumulation of silver in farm ponds and lakes where clouds have been seeded for several years. One such investigation showed that the amount of silver in pond water was less than that in a cup of coffee after it had been stirred with a silver spoon. Therefore there seems to be little basis for concern as might be anticipated by considering the extremely small amount of seeding agent added to the atmosphere in comparison to the large area involved.

Several states have operational weather modification programs, especially in the semiarid belt extending from Texas to North Dakota. These are funded by the states or counties involved and in some cases by federal agencies in conjunction with tests for evaluating the effectiveness of cloud seeding.

Instead of adding extra water, another possible option is to try to reduce the amount of water lost by vegetation (Figure 14-14). A growing crop uses an enormous amount of water, some of which seems to be unnecessary. One way to reduce this loss would be to cause the plants to close their stomata; the tiny pores in leaves where water escapes. Certain chemicals such as some of the ammonia compounds are effective in triggering this response and may reduce the amount of water lost. However, the plant grows because it takes carbon dioxide from the atmosphere and combines it with water to form plant tissue. If the carbon dioxide cannot get into the plant through the stomata, then the growth rate is curtailed.

Another alternative is to develop vegetation with a different color to reflect more radiation, such as white; the problem with this is that chlorophyll, which is necessary for photosynthesis, is green. Yet

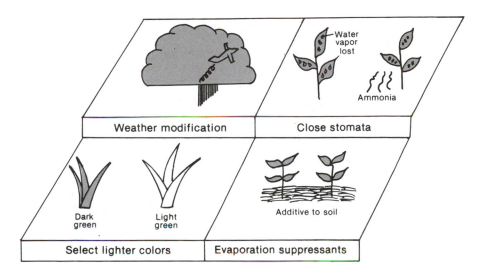

Figure 14-14. Different approaches to solving the water supply problem might be weather modification, or such other actions as selecting lighter colored vegetation, inducing the closing of stomata, or adding evapotranspiration suppressants to the soil.

another suggestion involves monomolecular films. Certain compounds such as some of the fatty alcohols will spread over a water surface to form a thin film only one molecule in thickness which retards evaporation. Some of these type compounds when mixed with water in the soil have reduced the evapotranspiration rate significantly. However, plant growth has been reduced accordingly, indicating that it may not be possible to separate large water usage from large yields.

Modification of Temperature

Another effort to modify weather for agricultural purposes involves air temperature. Frost protection of crops is a multimillion dollar business. One method in use is to take advantage of the latent heat of fusion of water, which is 335 J (80 cal) for every gram of water that freezes. As the air temperature approaches freezing, water sprinklers are turned on. As the water begins to freeze on the crop, the temperature stabilizes at $0°C$ because of the heat released as water freezes. This method is effective only until the ice load becomes too heavy to be supported by the vegetation. Thus, it is more useful for orange groves than for tomato plants.

Another approach to frost protection is to use wind machines to mix the air (Figure 14-15). Freezing temperatures usually occur at night under the influence of a surface inversion. Thus, the coldest air is near the ground and warmer air is above. If wind machines similar to an airplane propeller driven by a motor are used to stir the air, this brings the warmer air from above down to the ground. This mixing may be effective in preventing a killing frost.

Another method of frost protection is direct heating by means of burners or smudge pots in orchards. Placement of 40 to 50 heaters per acre may increase the temperature by 3°C. Combinations of heaters and wind machines are also used. As shown in Figure 14-16 the temperature is modified more by the use of both methods than by either one separately.

The temperature environment of vegetation can also be modified significantly by creating ridges in the surface of the soil. Ridges 15 cm high spaced 25 cm apart alter the temperature of the soil considerably, since ridge tops receive much more sunlight than furrows, and may

Figure 14-15. This wind machine is used for frost protection within an apple orchard near Lexington, Missouri. Such wind machines are effective when a strong surface temperature inversion is present.

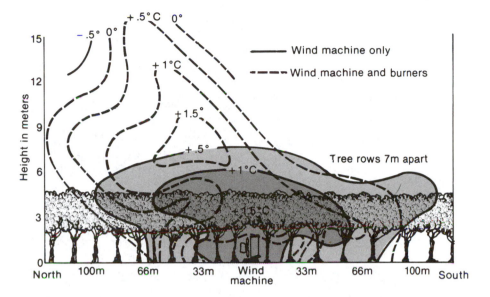

Figure 14-16. The temperature changes produced by a wind machine alone, and in combination with orchard burners are shown here. (After Crawford, 1965.)

also be drier. Measurements have shown temperatures within such ridges to be 15°C warmer during a sunny day than in the furrow between ridges. Warm weather crops can be planted earlier in the spring if planted in ridges since the warmer temperature will germinate the seeds.

Modification of Wind and Radiation

Shelter belts are sometimes used to improve the local climate for agricultural purposes and for the prevention of soil erosion by wind. Many shelter belts consisting of a row or several rows of trees were planted throughout the central United States after the severe dust storms of the 1930s (Figure 14-17). Such shelter belts have been shown to be effective in reducing the wind speeds. The wind speed is less than in the unbroken terrain for a distance of 20 shelter heights downwind. A shelter belt also influences radiation and moisture levels in the soil. Solar radiation is reduced on either side of a north–south shelter belt and this reduces the rate of moisture loss from the soil. Since plants need sunlight for growth the overall effect on crop yields

Figure 14-17. Shelter belts represent a method of modifying the microclimate for vegetation since the wind, radiation, and other climatic factors are influenced by shelter belts. (Courtesy USDA Soil Conservation Service.)

depends on whether the more plentiful soil moisture or reduced sunlight predominates. Some measurements have shown increased yields near shelter belts indicating the conserved soil moisture was more important.

Thus, a number of techniques are used to modify weather variables for agricultural purposes. It is clear that we are not able to control the weather and specify the optimum for food production or other purposes, but the modification techniques being used represent an important beginning in our attempts to change the amount of rainfall, the temperature of the air, the amount of radiation, and the wind speed for our benefit.

SUMMARY

Several inadvertent modifications of the urban climate are common. The amount of solar radiation at the surface within cities is decreased. The solar radiation is 15 to 20% less in urban areas than in surrounding rural areas; however, the temperature of cities is greater than in surrounding rural areas. Measurements of the temperature in cities of various sizes have shown that the heat island effect may increase from about $2°C$ to $4°C$ as a city grows from a population of fifty thousand to one million people. Accompanying the temperature increase is a decrease in relative humidity. Winds tend to blow toward the center of a city because of the warmer temperatures. Cities in general contain more pollutants than surrounding rural areas; although the air quality over cities has improved in the last several years as major efforts have been made to limit the pollution of our atmosphere. There is some indication that the amount of precipitation downwind from Chicago is increased. Investigations in other areas have revealed no large scale modification of the amount of precipitation because of the presence of cities.

Weather modification is practiced in some areas for agricultural purposes. Cloud seeding techniques are used in an effort to increase the water supply for agriculture. Silver iodide or dry ice is carried into the clouds with airplanes to supply them with an abundance of deposition nuclei. Warm clouds are sometimes seeded with salt to furnish condensation nuclei. Some experiments have shown an increase in precipitation of about 10% from seeding cold clouds.

Frost protection may be critical for orchards and other crops. Some frost protection is provided by the use of wind machines and heaters placed within an orchard. The temperature may be increased by a few degrees by these methods. Therefore, even though we are not able to control the weather completely, it is possible to manipulate it slightly to our benefit.

15
Laboratory Tornadoes

As I pulled on my leather gloves and carefully placed the slices of dry ice to cover a large section of the floor, I felt a heightened sense of anticipation. If my calculations were correct the crosswinds more than a meter above my head would interact perfectly with the rising airflow to duplicate the formation of one of natures most fascinating phenomena. As I flipped the switches to start the air currents and flood lights, a hush fell over the group that had gathered to witness the experiment. For one long minute nothing happened beneath the simulator. As we watched the floor, a couple of meters from our feet, the dry ice cloud showed a convergence of air into a peak that suddenly shot past our eyes and above our heads into the vortex generator. To the delight of everyone a tornadolike vortex was born and continued a twisting, swaying dance that seemed to be designed especially for the occasion. . .

ATMOSPHERIC VORTICES

Vortices form in the atmosphere every day. These vortices are swirling masses of air that can be as small as your finger or so large that they cover half the United States. One particular atmospheric vortex, the tornado, is generated almost 1000 times per year in the United States. A common characteristic of all vortices is that they are composed of air generally rotating horizontally about a vertical axis. The vertical core can be composed of an updraft, a downdraft, or it can contain no atmospheric motion.

The largest atmospheric vortices, other than the global winds, are the traveling frontal cyclones that represent a dominant feature of the weather of all midlatitude locations. The smallest visible atmospheric vortices are whirlwinds (dust devils) that can be seen on any hot summer afternoon in desert regions, and less frequently in other locations. While the dust devil and hurricane are surface generated atmospheric vortices, the large midlatitude cyclone and the tornado are generated by air currents above the surface and spiral their way downward.

We are now beginning to recognize that the large frontal cyclones are originated by the jet stream and extend down to the surface where they determine our weather on a particular day. It is now also recognized that tornadoes are generated within the interior air currents inside a thunderstorm several minutes before the circulation becomes strong enough to reach the ground as a tornado.

The difficulties of obtaining measurements within a tornado and within the generating region of a thunderstorm are obvious because of the high winds associated with a tornado. The recent use of dual Doppler radar has enabled us to see some of the air currents within a thunderstorm as it generates a tornado; however, radar is only sensitive to water droplets and does not see the air itself. Therefore, the winds of the tornado are still largely unmeasured, but it may be possible to model their development in the laboratory.

The nature of the atmosphere makes it difficult to analyze atmospheric storms in the laboratory as a chemist would analyze a solution in a test tube. Although it is not extremely difficult to generate a vortex in the laboratory, it has been more difficult to determine how a laboratory vortex is related to atmospheric vortices formed outside.

GENERATING LABORATORY VORTICES

A laboratory vortex that was created by placing a pan of heated water inside a glass cage with a circular opening at the top is shown in Figure 15-1. The panels of the glass cage all contain an open slot on the right-hand side. This allows the incoming air to start a spiraling flow as the air rises above the heated water and goes out through the top; thus, an atmospheric vortex is created, but the relationship between this vortex and a tornado or hurricane is very questionable.

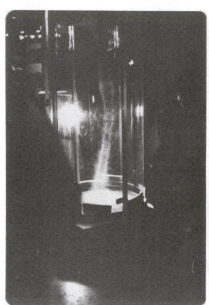

Figure 15-1. This laboratory vortex produced in a cylinder 2.4 m tall and 0.75 m wide, designed by the author for the 1980 Kansas State Fair, was effective in generating considerable interest. (Photographs by Greg Eagleman.)

A more complex enclosed vortex is created in the laboratory at Purdue University. A vortex is generated inside a cylinder 2.8 m wide and 2.7 m tall (Figure 15-2). Air is exhausted at the top of the cylinder, while incoming air at the base is forced into rotation by a rotating screen. Various vortex configurations are produced as the volume flow rate of air and rate of rotation of the screen are changed (Figure 15-3). As the rate of rotation increases sufficiently a single vortex breaks into multiple vortices.

An unconfined laboratory tornado was generated at the University of Kansas by simulating the larger circulation of the mesocyclonic vortex with a rotating cage and dynamic updraft in the center by creating a negative pressure with large vacuum tanks. The cage of 45 cm diameter was rotated at varying speeds while the pressure at the 5 cm diameter opening at the top center of the cage was varied from atmospheric pressure down to 0.1 atm. Different combinations of rotation and negative pressure produced tornadolike vortices of

Figure 15-2. Laboratory tornadoes are produced at Purdue University within a large rotating cylinder. (Courtesy of John T. Snow.)

different appearance (Figure 15-4). More rotation produced a larger vortex with more irregular edges while more suction and less rotation produced a smaller vortex with smoother outer edges and a well defined central core.

SIMULATING WINDS IN THUNDERSTORMS

Immediately prior to the development of dual Doppler radar, a theoretical model, the double vortex thunderstorm model, was developed from mathematical equations to explain the winds required in a thunderstorm to generate a tornado. Thunderstorms represent a barrier to the surrounding airflow and the application of the equations describing airflow around a barrier reveal that a double vortex would

Figure 15-3. Examples of a laboratory vortex and multiple vortices produced in the laboratory at Purdue University. (Courtesy of John T. Snow.)

be generated within the thunderstorm to resist the surrounding airflow. The double vortex consists of counterclockwise circulation around a vertical axis in the whole southern half of a thunderstorm and clockwise rotation around a vertical axis in the whole northern half of a thunderstorm moving toward the northeast.

Tornadoes are known to originate in the southern part of a thunderstorm as shown in Figure 6-1. The counter rotating vortices within the thunderstorm are intensified by the crosswinds of the atmosphere

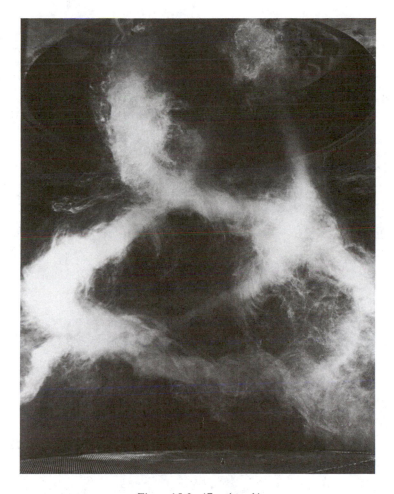

Figure 15-3. (Continued.)

composed of west winds in the upper and middle parts of a thunderstorm and south winds in the lower part of a thunderstorm moving toward the northeast. The moving thunderstorm increases the inflow of low level air from the east and intensifies the opposing nature of the air currents.

Recent dual Doppler radar data has verified the existence of a double vortex in severe thunderstorms. Thus, the structure of a tornado producing thunderstorm is very organized with an updraft and

Figure 15-4. An unconfined laboratory vortex can be generated by a combination of rotating air produced by a rotating screen with a suction up through to the center of rotation. The visible material in this case was dry ice.

crosswinds in the southern part of the thunderstorm that generate the tornado. This mechanism is beginning to be known as the mechanical theory for the formation of tornadoes.

Such information on the wind structure within a tornado producing thunderstorm can be used to simulate these same opposing air currents in a laboratory to produce a laboratory tornado.

LABORATORY TORNADOES PRODUCED BY BLOCKAGE

Since the flow of air around a barrier produces vortices behind the blocked air current, it should be possible to combine this circulation around a barrier with an updraft to produce a laboratory tornado as shown in Figure 15-5A. I have generated such a vortex by using a 3/4 hp electric motor with a four-bladed propeller operating at 1,725 rpm to produce an updraft, and ordinary 51 cm (20 in) window fans to simulate the horizontal winds around a curved metal shield that formed a barrier. These were mounted 2.5 m above the floor to produce a vortex as shown in Figure 15-6. The laboratory tornado extends down to a table located 1.5 m below the curved metal shield in this photograph, or down to the laboratory floor. The vortex is made visible by placing a layer of dry ice over the table or floor. The central core of the vortex is visible in Figure 15-6 and has an appearance very similar to some photographs of tornadoes and waterspouts. Water spouts more frequently show the central core since they are generated by thunderstorms over oceans or lakes where dust and debris does not hide the vortex structure (Figure 15-7). This water spout formed near the Florida Keys and clearly shows the central core with rotating winds around it.

LABORATORY TORNADOES FROM CROSSWINDS

Since the thunderstorm is able to generate a tornado by horizontal crosswinds combined with an updraft, it should be possible to generate a tornado-like vortex in the laboratory by these same air currents. The two smaller fans were aimed in opposite directions to simulate the opposing horizontal winds, while the third fan simulated the thunderstorm updraft as shown in Figure 15-5B. The updraft was narrowed through a circular opening of 20 cm diameter. A photograph of a laboratory tornado produced by this arrangement is shown

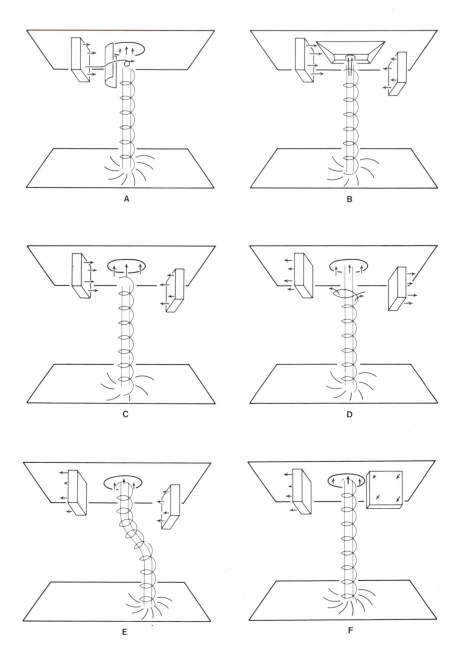

Figure 15-5. The most realistic laboratory vortices should be produced by simulating the air currents within thunderstorms. Several air currents used to create a tornadolike vortex in the author's laboratory are shown here.

Figure 15-6. An air current blowing around a curved metal shield will produce a vortex if there is an updraft at the top of the shield.

in Figure 15-8 as it extends from near the ceiling down to a table about 1.5 m below.

It was found that larger and stronger laboratory tornadoes were generated by making the updraft opening as large as the blades of the fan. Figure 15-9 shows a vortex generated at the variable-height mounting board located 2.5 m high that extended down to the floor of the laboratory. The opening of the updraft area was 55 cm with the two 51 cm diameter fans supplying crosscurrents to simulate the horizontal winds in the tornado generation region of the thunderstorm as shown in Figure 15-10.

VARIATIONS IN CROSSWINDS

A number of interesting variations in the direction and speed of the crosswinds were tested with the laboratory tornado. Figure 15-5C

Figure 15-7. This water spout shows shells of circulation around the central core very similar to the laboratory vortex. Some tornado photographs also show this appearance. (Courtesy of J. H. Golden, NOAA.)

shows a combination of crosswinds and updraft that produces a weak intermittent vortex between the floor and the mounting board positioned at a height of 2.5 m. The opposing crosscurrents in a horizontal direction feed the rotating air in the upper part of the vortex. The vortex forms, disappears, and then reforms as the three air currents interact in different ways.

Another combination of crosswinds that produces an intermittent vortex is shown in Figure 15-5D. In this case the horizontal crosswinds are aimed outward from the updraft area in an effort to pull the outer rotating part of the vortex upward instead of feeding it as in the previous figure. The vortex produced by these air currents is intermittent since the vortex is destroyed as the outer shell is pulled away from the central core. This rapidly reduces the strength of the vortex much like the effect of figure skaters extending their arms outward to slow their rate of spin. Thus, the laboratory tornado repeatedly forms and is destroyed by this effect.

Figure 15-8. When the opening for the updraft is decreased, a vortex is still produced but its diameter is slightly decreased.

Another interesting arrangement of air current was produced by removing one of the horizontal air currents shown in Figure 15-5D to produce an intermittent laboratory vortex. The vortex was generated by the updraft and a single horizontal air current blowing away from the updraft area. This particular arrangement is interesting since it shows that the main requirement for generating an atmospheric vortex is a deficiency of air. As the air rushes in to fill the partial vacuum and begins a spiraling pattern the rate of air flow into the area of lower pressure is reduced and this allows the whole process of filling the partial vacuum to be prolonged.

Figure 15-9. The diameter of the full length laboratory vortex can be decreased in particular regions by interfering slightly with the airflow surrounding it.

One of the largest laboratory tornado-like vortices is generated by the somewhat surprising arrangement of air currents shown in 15-5E. This laboratory tornado is more stable and not as intermittent as those produced by the previous arrangements of air currents. It is generated by the updraft with both the horizontal air currents going in the same direction. It is not obvious why this combination of air currents should create a stronger vortex than the arrangement of air currents previously shown, but it indicates that wind shear may be an important parameter.

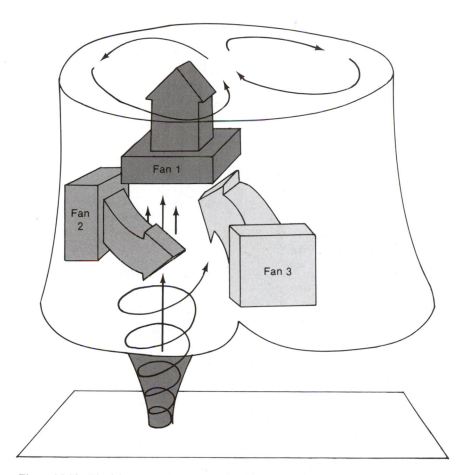

Figure 15-10. The laboratory tornado is generated by simulating the updraft within a thunderstorm and the opposing air currents as shown here for a thunderstorm with a tornado in the southern part.

A steady and persistent laboratory tornado is produced by the arrangement of air currents shown in Figure 15-5F, where the cross-currents blow on different sides of the updraft at a 90° angle to each other. This creates a strong vortex between the ceiling mounting board and the floor of the laboratory. It is interesting to note that these air currents are somewhat similar to the winds in the atmosphere during the time of tornado formation, with southeast surface winds going into the central part of the thunderstorm as the winds on the west side of the thunderstorm flow from the northwest around the back side of the thunderstorm.

OTHER OBSERVATIONS OF LABORATORY TORNADOES

A very interesting characteristic of the laboratory vortex is shown in Figure 15-11. After the vortex has been generated by an updraft and horizontal air currents, the vortex may attach itself to one of the side fans. It soon "winds down" in this position, however, since it has lost the source of energy for rotating air. This characteristic probably applies to other atmospheric vortices, such as tornadoes, and

Figure 15-11. After its generation, the vortex may attach to a side fan instead of the central updraft. This causes it to decay rapidly as indicated by the lower part of the vortex.

indicates that they would attach themselves to the nearest region with a net outflow of air.

The vortex is composed of several shells of circulation as shown in Figure 15-12. The outer shell can be stripped away leaving the central stronger circulation around the core itself. A destruction of this outer shell leads to more rapid decay of the whole vortex. The

Figure 15-12. The various shells of rotation can be seen in the lower part of this laboratory vortex. The central core and smooth vortex are surrounded by a shell with more irregular circulation.

last stage of a tornado is frequently a rope stage as shown in Figure 15-13. This may mean that such factors as surface interaction with the tornado, interaction between the tornado and the rain area, or improper air currents within the thunderstorm have been responsible for reducing the outer circulation of the tornado and contributing to its decay.

Figure 15-13. Only the inner core and smooth surrounding shell are visible in this rope-shaped laboratory tornado.

Surface interactions can be investigated with laboratory vortices as shown in Figure 15-14. Some scientists in the past have suggested a downdraft within the core of a tornado while others have felt that an updraft should exist there. Direct measurements in a tornado are, of course, not available. But, the confined laboratory tornado clearly shows that foam peaks upward into the core. This is consistent with lower pressure and an updraft through the core of the vortex.

If a traveling tornado is simulated by moving a sheet of plywood beneath the unconfined laboratory tornado it reveals airflow patterns that compliment and amplify tornado damage investigations. The path of the vortex across a viscous liquid coating on the plywood reveals airflow primarily in the direction of travel of the vortex (Figure 15-15). Slow movement produces a looping path while a faster rate of tornado travel produces a straight path.

Another interesting observation of the laboratory vortex is that after it has been generated, it exists because of its own momentum for some time after the generation mechanism has been shut down as

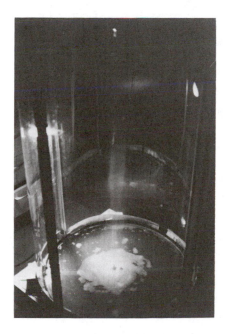

Figure 15-14. The inverted barometer effect is produced by the confined laboratory tornado. Low pressure inside the core of the vortex causes foam to form a peak (left) that rises upward until it collapses (right) or is broken and centrifuged out of the vortex.

Figure 15-15. Path of the laboratory vortex as its motion was left to right across a viscous fluid applied with up and down brush strokes. Major airflow within the base of the vortex was left to right in the same direction as vortex movement.

shown in Figure 15-16. This illustrates that the air has momentum that allows it to maintain its velocity after it has been developed. Thus, the tornado may last for a short time after the generation mechanism has been eliminated.

APPLICATIONS

Since the laboratory vortex was generated by simulating some of the air currents within thunderstorms as developed theoretically and measured by dual Doppler radar, it is helpful in understanding some of the characteristics of tornadoes. One of the conclusions of these experiments with a laboratory vortex is that tornadoes can probably be generated by a variety of air currents within the atmosphere. This is undoubtedly one of the reasons why the smaller tornadoes are so difficult to forecast. If they can be generated by such a variety of crosswinds then the only hope in being able to forecast small tornadoes is to obtain information on the intensity of the thunderstorm updraft.

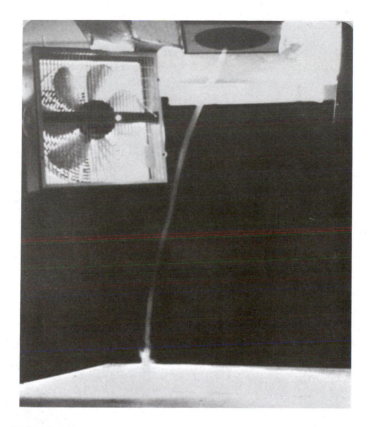

Figure 15-16. Laboratory vortex remaining in existence after the generating mechanisms have been stopped.

The updraft must be large enough to create a deficiency of air that will support a vortex from the central part or lower part of the cloud all the way down to the ground. Larger tornadoes are more easily forecast as they correspond to a more organized structure within the thunderstorm that consists of large counterrotating vortices. Such thunderstorms produce tornadoes that last longer and are more damaging.

One of the reasons for studying tornadoes in detail is to determine their particular formation mechanism and structure, since such an understanding may make it possible to consider controlling them sometime in the future. However, we have no operational programs attempting efforts in this direction based upon our current level of understanding of these atmospheric storms.

SUMMARY

Atmospheric vortices can be generated in the laboratory to simulate some of those that occur in the atmosphere. In general, vortices that are enclosed within cylinders or cages are not as realistic as unconfined vortices. One of the most realistic tornado-like vortices is generated by simulating the winds in a thunderstorm. These include opposing air currents in the presence of a large updraft. It can be shown in the laboratory that flow of air around a barrier will produce vortices behind the blocked air current.

Various combinations of crosswinds can be investigated for producing laboratory vortices. Several different combinations of crosswinds will interact with an updraft to produce a vortex. A persistent laboratory vortex extending from the ceiling of the laboratory to the floor was produced by air currents blown horizontally at an angle of 90° to each other, in combination with an updraft these produced a strong laboratory vortex. Laboratory vortices can be used to study the detail structure of a vortex, including its various component parts such as the updraft core and shells of circulation with various speeds around this core. In addition to the generation mechanism, a laboratory vortex can be used to study the air currents as the vortex interacts with the surface beneath it.

16
Softening the Blow

There was very little warning that this thunderstorm was any different from the hundreds of others that had passed over our house. At first I dismissed the rising noise level outside as a low flying jet, but realized that a tornado was upon us as I saw the neighbors roof lift, spin, and suddenly dissappear. As we tried to rush to the basement the windows shattered and we hit the floor. The next thing we knew we were suspended in air and the floor was ripped from beneath us. We were thrown against some bushes where the northeast part of our house had been. Our only thoughts were for our own safety and not of the magnitude of our loss at that moment . . . .

DISTRIBUTION OF TORNADO DEATHS

The distribution of tornadoes is such that more tornadoes occur in the Central United States than any other area, comprising a band of greatest activity through Oklahoma, Kansas and Missouri as previously shown in Figure 6-5. If tornado activity is combined with the number of people, the potential for tornado deaths should result. This produces a band of high risk in the Central United States because of many tornadoes and in the eastern United States because of the high density of population. However, observed tornado deaths are more concentrated in the southern United States, centering in Mississippi, Alabama, Arkansas, Louisiana, and Texas (Figure 6-10). The design and type of houses, lack of basements, poor information concerning protection during tornadoes, and less concern about tornadoes may account for the high number of deaths in these areas. Since it is impossible to control tornadoes it is important to have good information on designing houses and seeking protection during tornadoes, based on past observed effects of tornadoes.

TOPEKA TORNADO DAMAGE

The Topeka tornado was very large, measuring about 1 km in diameter at the ground (Figure 16-1). It produced almost complete destruction in its path. Destruction of houses was produced by high wind speeds, related pressure changes, and debris. Boards, tree limbs, roofs, and other debris were carried in winds with speeds of several hundred kilometers per hour. Such debris punctured the south and west walls of buildings. Looking to the southwest from Topeka is Burnett's Mound which was supposed to protect the city, according to Indian legends (Figure 16-2).

After the tornado came directly over Burnett's Mound it hit a residential section in Topeka composed of houses with walkout

Figure 16-1. The Topeka Tornado on June 8, 1966, was quite large, about 1 km in diameter at the ground.

Figure 16-2. Burnett's Mound, to the southwest of Topeka is shown in the background behind houses severely damaged by the tornado.

basements that all faced the southwest. The northeast side of the walkout basements were mostly underground, extending only about 1 m above ground. The walkout side on the southwest was more exposed to tornadic winds. The southwest rooms, in general, were destroyed, but the floor covering the poured concrete basements stayed in place in a high percentage of the cases (Figure 16-3). This was one reason why more people were not killed in this section. In all, 16 people were killed by this tornado.

A number of "survival" stories circulated from this section of the city after the tornado, including the children who hid under a pool table in the southwest rooms, the two couples who hid in a bathtub with a mattress pulled over them on the main floor above the basement, and a party of people with one brave but stubborn woman who insisted on staying upstairs while the tornado passed. She was found hanging over a wall remnant, unhurt, after the tornado. Figure 16-4 shows one of these houses that was penetrated by debris and demolished by the high winds.

Figure 16-3. Walk-out basements facing the southwest generally had a part of the floor remaining over the northeastern section.

Figure 16-4. The southwest sides of the walk-out basements in Topeka were severely damaged by the tornado.

Observation of damaged houses showed that the larger rooms were less safe than the smaller rooms. One of the reasons for this is the added strength from the additional walls per unit area. The closets of a room are frequently left standing while the walls of the larger rooms are blown away by the tornado.

The next section of houses in the path of the tornado was composed of multiple family brick houses. These were apartment houses holding four families. Each house has a basement, mostly underground, and two stories above. The top story was the most unsafe while the basements were safest. Basements were unsafe only along the south sides where debris came through parts of the aboveground walls and through the windows (Figure 16-5).

Hundreds of single family brick and wood frame houses were damaged or destroyed by the tornado. The brick walls of the houses were crumbled by the high winds in a manner similar to the frame houses. On the southwest corner of several blocks, houses were shoved to the north by the tornado coming from the southwest and piled together down the street (Figure 16-6). One elementary school constructed of brick was in the tornado path. It lost the roof and part of the walls especially those on the south and west sides.

Figure 16-5. The south side of this multi-apartment house shows the amount of debris that bombarded and piled up against it.

Figure 16-6. Houses located on the southwestern corner of a block were frequently jammed against each other toward the north, as shown in this photograph.

Washburn University campus was badly damaged by the tornado (Figure 16-7). Most of the buildings were built from stones held in place by mortar, and these were badly crumbled by high winds. Most of the buildings had to be torn down and rebuilt as a result of the damage. MacVicar Chapel held more than 50 people who were attending a musical recital during the tornado. They rushed downstairs as the tornado approached and sought refuge in the southeast part of the basement which turned out to be the only possible place in which they could have survived. The southwest room was packed with large stones from the upper parts of the walls as they came crashing through the wood floors to pile up past the level of the ceiling of the lower rooms.

The next residential section contained large two-story homes. Many had weak foundations which allowed the whole house to be blown toward the northeast. Some of the houses were turned upside down as shown in Figure 16-8. Along one street more than ten of these houses lost their south porch to the tornado. Some of them also lost roofs (Figure 16-9).

Figure 16-7. MacVicar Chapel on the Washburn University campus where 50 people sought shelter from the tornado.

Figure 16-8. Older houses with weak foundations frequently were turned upside down as the one in the central part of this photograph. The house on the left has been shifted from its previous location up from the stairs.

Figure 16-9. South porches were uniformly removed from many houses by the Topeka Tornado.

Even if the framework is retained, the interior of the buildings may be severely damaged by pressure effects. Pressure damage may be severe due to the lower pressure inside the tornado as it passes over a house. Such differential pressure can cause a building to virtually explode. The greater internal pressure in houses has the beneficial effect of sending walls of houses outward instead of inward on top of persons located inside. Only south or west walls were occasionally blown inward by the stronger winds blowing in the direction that the tornado traveled.

In the downtown area, the tornado was wide enough to hit a 10-story building and a water tower which were separated from each other by a few blocks. The water tower was undamaged but the office building was gutted. All the windows were broken, the shades battered, and much of the framework was ripped away. The reinforced concrete structure of the office building remained undamaged. The tornado passed close enough to the State Capitol Building to pull out one of its copper panels from the dome.

The next residential section also contained two-story houses with weak foundations. Houses were moved downwind, turned upside down, and generally suffered extensive damage. Houses with poured concrete foundations would be expected to retain the floor above, but many did not because of either no bolts in the concrete or no nuts on the bolts. One house with a brick foundation was shoved to the northeast a few meters. Much debris entered in the south and west sides of the basement, whereas the house slid over the ground in the northeast section which resulted in much less danger to those who could have sought shelter there. A similar example is shown in Figure 16-10. The concrete block foundation failed and the whole house was pushed to the north. Sixteen people were in the southwest corner of the house where concrete blocks, boards, and pipes flew in. Fortunately, no one was badly hurt. In the northern part of the house, no debris had fallen in.

Damage statistics collected from houses with basements damaged by the tornado in Topeka showed that the south and southwest parts were most dangerous with about 50% of them unsafe, while only 25%

Figure 16-10. This house was pushed toward the northeast by the Topeka Tornado allowing the southern part of the house to drop into the basement. Sixteen people sought shelter in the southwest corner of this basement.

were unsafe in the northeast part of the submerged basements (Figure 16-11). The reasons for the unsafe locations included weak foundations, movement of the house in the direction of the tornado, and debris coming through the south and west sides of the structure.

Analysis of the walkout basements oriented with the southwest wall toward the tornado revealed that only 18% of them were unsafe in the northern parts, but 88% were unsafe in the southern rooms.

Damage statistics for the first floors of houses damaged by the tornado showed that the most unsafe areas were along the southwest, south, and southeast. The safest locations in the houses were in the northern sections where only 30% were unsafe compared to 50% unsafe in the southern rooms.

In 1948, a Texas tornado virtually destroyed an amateur weather observer's house. The observer was hanging on to a bedpost looking up into the vortex. He described the funnel as being very smooth inside with brilliant lightning displays. He saw a house next door disintegrate the second the leading edge of the tornado touched it. The tornado was about 7 m off the ground as the trailing edge of the vortex approached and he escaped further damage. This fits other tornado damage and laboratory vortex observations that indicate much of the major damage was from winds going in the direction of movement of the tornado or slightly to the right when looking in the direction the tornado traveled. This means that the inflow from the rear of the tornado provides the mass of air to be accelerated to velocities capable of causing damage (Figure 16-12). This also explains the

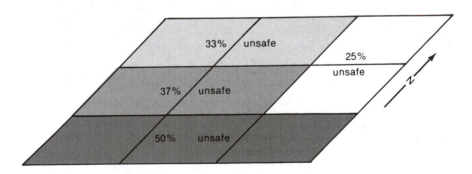

Figure 16-11. Damage statistics gathered after the Topeka Tornado showed that the most unsafe locations within houses were those along the south, and the safest locations were those opposite the approach direction of the tornado or the northeast part of basements.

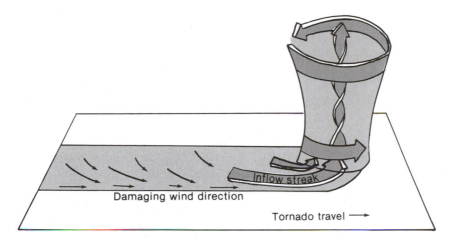

Figure 16-12. The most damaging winds are frequently in the direction that the tornado travels. This is because of the inflow of air from behind the vortex.

observed debris punctures almost exclusively in the south and west sides of a house struck by a tornado coming from the southwest (Figure 16-13).

LUBBOCK TORNADO DAMAGE

The Lubbock tornado was the next single tornado to cause over $100 million damages. The Lubbock tornado formed from a very large thunderstorm just after sunset on May 11, 1970. The winds in this tornado were so strong that houses were completely destroyed and trees were uprooted or their bark was peeled off by the winds. Asphalt roofing from houses was forced into the dry ground so deep that it could not be pulled out (Figure 16-14). Metal roofing was wrapped around posts as if it were cloth. Large utility poles were ripped out or splintered. Much debris landed in a field east of Lubbock outside the damage area.

The tornado tore through a residential section containing some of the finest houses in Lubbock. Basements are rarely built in this city or others in the South. Families sought shelter within their houses on the first floor or in some cases in an outdoor storm shelter located in their yards. It is amazing that the tornado killed only 26 people as it destroyed hundreds of houses. Boards were driven

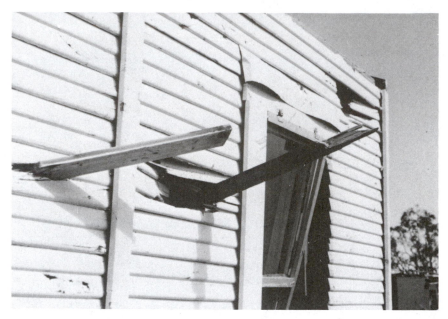

Figure 16-13. Boards punctured the southern wall of this house. Such penetration by debris was common on the south and west walls and very uncommon on the east and north walls.

Figure 16-14. Asphalt shingles from the roofs of houses penetrated the soil with such force that they could not be pulled out after the Lubbock Tornado of May 11, 1970.

through walls into rooms inside, particularly on the south and west sides of houses. Metal was stripped away or bent on vehicles that apparently were not moved far (Figure 16-15). Houses with strong wall-to-concrete foundation connection were ripped off at a height of a meter or so above the ground (Figure 16-16). In some cases the roof was lifted off and the walls fell outward. For some families, this tornado was the second or third that had destroyed their homes; most were rebuilding in the same location.

An aerial view shows how some of the roofs were damaged first at the peaks and eaves, then as a whole. Progressively greater damage is shown in Figure 16-17. Even in these well-constructed houses the roof-to-wall and wall-to-foundation connections show up as weak parts of a house.

Statistics collected on safest locations on the first floor of houses showed that 76% of them were unsafe in the southwest and south. The safest places were the central, north, and northeast rooms. Similar investigations following other tornadoes have also shown that the previously popularized southwest part of a house is the most unsafe during a tornado. Safest locations are these on the opposite side of the house from the approaching tornado (Figure 16-18).

DESIGNING WIND RESISTANT HOUSES

The ability of houses to withstand high wind velocities can be greatly improved by proper planning and building design. More specifically, orientation and roof slope, quality of construction materials, and strength of connections of members greatly affect the strength of a house. Houses can be modeled and placed in a wind tunnel to obtain measurements during winds of various velocities. Models of houses were constructed and pressure measurements were obtained as the winds increased to more than 300 km/h.

The angle of the roof was an important factor in the destruction of model houses by strong winds. Low angle roofs are lifted off from the same uplift forces that cause airplanes to fly. Air flowing over a 15° roof increases in speed with a resulting decrease in pressure that may be sufficient to lift the roof unless it is anchored firmly to the walls. This can be accomplished by more nails or by "hurricane clips," metal plates with teeth punched out to bite into wood after they are nailed in place.

Figure 16-15. An indication of the extreme winds within the tornado can be gained from these photographs. These vehicles do not appear to have been rolled; the metal was simply bent by the strong winds.

Figure 16-16. Houses such as this were very well constructed with strong wall-to-foundation joints. The winds, however, were so strong that they took the roofs and half the walls.

Another way to improve the ability of a roof to withstand winds is to use a steeper roof design. A roof with a 45° angle is not exposed to the reduced pressure from airflow over it since it is too steep for the "airplane wing effect." A roof with a 30° slope is much better at withstanding strong winds than one of lower angle (Figure 16-19).

Every roof design has a critical wind angle, the orientation that produces the greatest pressure reduction over the roof. The 15° roof on a rectangular house has much greater pressure reduction, and hence more damage, if the winds are perpendicular to the long side. The best orientation for withstanding strong winds is with winds against the short side and parallel to the long side of the house. Thus turning the short side of a rectangular house with a low-angle roof toward the southwest would improve the chances for withstanding strong winds.

A flat roof is subjected to considerable negative pressure on the roof at the most upwind location. This develops as the airflow is directed upward over the edge of the roof, and turbulence is created there. The negative pressure thus created makes a flat roof more susceptible to wind damage than a 30 or 45° roof.

We also investigated the effects of other parts of a house, such as porches and chimneys, on wind damage. Models were prepared and

Figure 16-17. Progressive damage is shown in this series of photographs. These range from the least damaged in the upper left to the most severely damaged in the lower right.

tested with results that showed such modifications were helpful in reducing damage. The airflow was modified by the obstructions in a beneficial way for all roof angles.

Another test involved differences between one- and two-story houses. This showed only minor differences between the two designs if they had the same roof angles. For each design the two-story house was damaged slightly sooner than the one-story house.

The patterns of failure were also of interest. The first parts of a house to be damaged as the winds increased were the eaves of the roof on the upwind side of the house and the downwind side of the

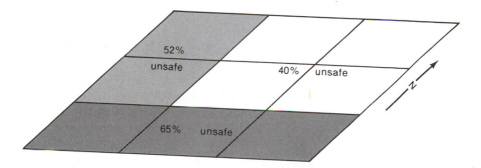

Figure 16-18. A composite of damage statistics gained from several different tornadoes has shown that the locations within houses that are most unsafe are those facing the approach direction of a tornado. Safer locations are opposite the approach direction of a tornado; since this is typically from the southwest, the northeastern part of a house is generally safer.

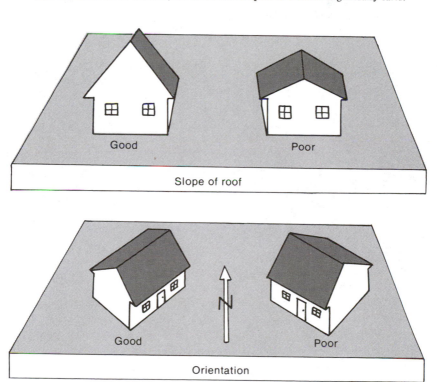

Figure 16-19. The design of a house contributes to its ability to withstand strong winds. Steeper slopes and houses with the short end oriented toward the southwest are more likely to withstand high winds.

peak of the roof. These structural damages began at about 250 km/h. As part of the roof blew away, the potential for additional structural damage increased. A model building destruction sequence is shown in Figure 16-20.

In summary, wind damage is most likely to occur differentially according to roof angle and orientation. Steep roofs are safer in strong winds than low roof angles. Wind damage is most likely at the upwind edges and along the peaks, at the roof-wall joint or the wall-foundation joint. The downwind part of a house is likely to receive less damage than the upwind section.

The southwest part of the first floor is unsafe because of the greater chance of structural damage in this part of the house, in addition to the greater chance for penetration of debris into this area. For houses with weak foundations, the southwest part is unsafe because of the tendency of tornadic winds to move the house off its foundation with displacement toward the northeast. Houses with

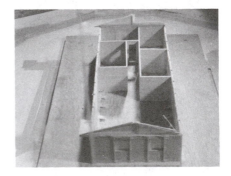

Figure 16-20. This series of photographs shows progressive damage during a destruction model test. Damage occurred first at the peak of the roof, then to the roof itself.

good foundations—poured reinforced concrete—are also the most unsafe in the southwest part because of the increased risk of the house and ceiling over the basement being removed from over the southwest room before those in other locations.

Houses with vents were tested to determine the effect of ventilation on damage. In general, vents in locations comparable to doors and windows on a single story house caused it to receive greater damage if venting was provided on either the upwind or downwind side. However, vents on the downwind side near the peak of the roof allowed the house to withstand strong winds.

A number of applications can be made that will allow houses to withstand stronger winds. In addition to the importance of slope and orientation of the house the proper strengthening of joints will allow a house to survive much stronger winds. Strengthening of the roof to walls by hurricane clips has already been mentioned. The walls should be securely anchored to the concrete foundation with bolts embedded in the concrete. Venting of the roof should be provided near the peak. This can serve the added purpose of releasing hot air from the attic during the summer.

While houses with poured reinforced concrete foundations with good wall to foundation connections are generally safe shelters from a tornado, additional security is provided by constructing a storm shelter as a part of the basement. Such an indoor shelter is preferable to an outdoor underground shelter because of the danger in getting to it during the short warning time for most tornadoes. The shelter should consist of concrete walls eight inches thick with a top slab of eight inches of concrete reinforced with metal bars connecting the walls and ceiling at nine inch intervals. A hallway with a 90° turn or strong door should form the entrance. This should prevent flying debris from entering the shelter. If such a shelter is constructed when the house is built it represents a very small additional investment.

SUMMARY

Since it is currently not possible to control or eliminate tornadoes, it is important to have proper information on seeking protection from tornadoes and on designing houses to withstand such high winds.

Observations of tornado damaged houses have shown that smaller rooms are in general safer than larger rooms. In general, the southwest parts of houses are the most unsafe for a tornado coming from the southwest. Since this is the typical approach direction of tornadoes, the northeastern part of basements has been determined to be the safest from damage investigations. The reasons for this include the observed pattern of destruction where houses with weak foundations were moved several feet in the direction the tornado was going. This allowed debris and parts of the house to fall into the southwest part of the basement. With poured concrete reinforced with metal, the floor ordinarily remained over the basement. If part of it was removed by the tornado it more frequently was the southwestern part of the floor. For these reasons the northeastern part of basements is usually safest.

If basements are not available, the northeast part of the first floor is also generally the safest. The reason for this is that a large amount of debris is carried by high winds within a tornado. Such debris bombard the south and west walls sometimes penetrating through them, while no similar destruction of the east and north walls occurs. In addition, if any walls are blown inward, they are generally the south and west walls because of the strong winds blowing in the direction the tornado is going.

Houses can be designed to better withstand wind forces. The angle of the roof is an important factor in determining whether a house can withstand strong winds. A steep roof, with a 45° angle will withstand much higher wind speeds than a low angle roof of only 15° or a flat roof. The flat roof or low angle roof is subjected to strong lifting forces in the same way as an airplane wing.

The orientation of a house can improve its ability to withstand strong winds. The best orientation for a rectangular house with a low roof angle is with the short side turned toward the strongest winds; this means that the short side should be facing the southwest to improve the house's chance of withstanding the strong winds associated with a tornado. The orientation of houses with steeper roofs such as 45° is not so critical.

Houses can also be strengthened to withstand stronger winds by making the joints stronger at the roof-to-wall connections and the wall-to-foundation connection. This can be done with more bolts in the foundation or by using more nails at the other joints. Thus, it is possible to improve the ability of a house to withstand the strong winds associated with a tornado.

17
Harvesting the Wind and Sunlight

I watched as sunlight penetrated through the almost transparent atmosphere to heat the land which in turn heated the lower atmosphere until it "boiled." The boiling air became visible as moisture condensed into a thunderhead that provided a dramatic manifestation of energy; free energy, ripe for the harvest. Beneath the thunderstorm, a tornado churned as it released the tremendous energy it obtained from the thunderstorm. If we built a giant vortex generator, perhaps it could give us continuous power extracted from sunlight and air currents intensified in an artificial tornado. This certainly would be "unusual weather" if we could create it, and then harvest the power from it...

ATMOSPHERIC AND OTHER ENERGY SOURCES

Energy projections (Figure 17-1) show little change in the production of electricity by water resources, while nuclear energy is projected to play a larger role in the production of electricity. Geothermal energy will be used in some locations, but the projected amount of energy from this source is not large. The total United States energy supply will continue to be provided for largely by oil, with it supplying almost half the total energy supply, but other sources, such as coal, gas, and nuclear, wind, and solar energy, will also be important.

Since our energy is projected to remain in short supply, alternate sources may become increasingly important for running homes, operating industry, etc. Important possibilities include harnessing solar and wind power.

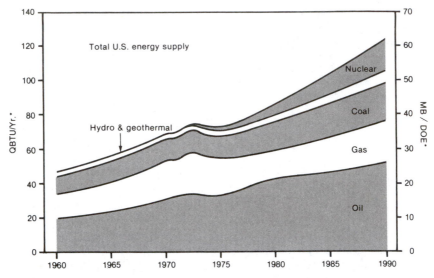

Figure 17-1. Projections for various energy sources are shown here to 1990. The scale on the left is in quadrillion British Thermal Units per year, while the scale on the right is in million barrels per day oil equivalent. (After EXXON Corporation.)

The energy involved in atmospheric processes is tremendous. For example, a thunderstorm 10^7 m^2 contains about 10^5 megawatts of power (Figure 17-2). This is approximately the power utilization of the entire state of California. A forest fire of the same size also has about 10^5 megawatts associated with it. The tornado has about 10^2 megawatts within its small size of 1000 or less square meters.

Within a thunderstorm, another source of energy is the electric potential that develops as a thunderstorm generates a charge separation between the ground and the base of the thunderstorm. It is estimated that the charge buildup before lightning strikes may be more than 100 million volts. Surprisingly, the amount of charge that flows may be only about 20 coulombs or enough to run an ordinary lightbulb for a little over a minute.

Another potential source of energy is the updraft in thunderstorms. An updraft 1 km^2 of 10 m/sec would contain 1000 megawatts of power. A further source of energy is the condensing water vapor. The latent heat of condensation of water vapor in an ordinary thunderstorm amounts to more than 10^5 megawatts. In comparison, a large nuclear power plant may produce 1000 megawatts and an ordinary power plant may produce 100 megawatts. Therefore,

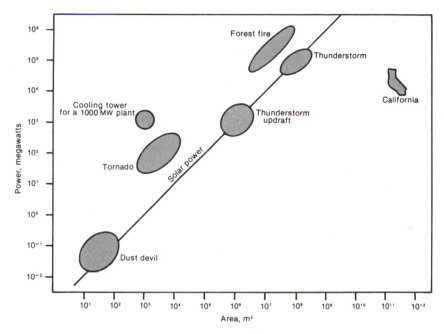

Figure 17-2. A comparison of the amount of power involved in various atmospheric phenomenon is shown here. (After Slinn, *Bulletin of the American Meteorological Society*, 1975.)

the potential for energy derived from thunderstorms is great, but we are lacking good suggestions for tapping this pollution free energy source.

One suggestion for harnessing the energy of the atmosphere is to use the sea breeze. During the daytime, land next to water heats and causes localized onshore breezes. It might be possible to utilize these winds with a huge plastic tube 10 km in diameter at the surface and 100 m at its top. The restricted diameter of the tube would cause the wind speed to increase. If we placed a generator in the region of highest air speeds it could generate electricity. This idea is probably not practical, however. If it were possible to build such a large tube, its appearance might prevent its use, in addition to its large cost and possible aircraft hazard.

Another "far-out" suggestion involves the use of waste products from a nuclear power plant. Waste heat is normally fed into reservoirs or cooling ponds to cool the water for discharge or reuse. Since warm, moist air is needed for thunderstorm formation, the cooling ponds could be surrounded by a few kilometers of asphalt with heating pipes for hot water to warm the asphalt near the surface. This should produce thunderstorms over the cooling ponds to help cool the waste water (Figure 17-3). In the right location, the extra water generated by the thunderstorm could be used to run a hydroelectric plant downstream. Cloudseeding techniques could also be used to develop larger thunderstorms.

WIND POWER

More conventional wind systems have the potential for supplying a significant proportion of our future energy requirements. It has been estimated that the total amount of available energy from the wind is equivalent to 1000 typical power plants of 1000 megawatts each. In comparison, this is about 10 times greater than the total potential for hydroelectric plants.

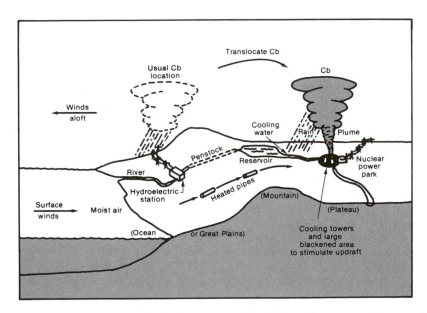

Figure 17-3. It might be possible to use the heat generated by nuclear power plants to redistribute thunderstorms if several factors work together. (After Slinn, *Bulletin of the American Meteorological Society,* 1975.)

The largest aerogenerator was built near Rutland, Vermont, in 1941 by the Central Vermont Public Service Corporation. The tower height was 32 m, and the diameter of the blades 53 m. It was operated successfully for 23 days, producing 1.25 megawatts of AC current at 13.4 m/sec wind. War shortages hampered the operation of the project and it was abandoned.

A current aerogenerator has been constructed with NSF and NASA funds at the NASA experimental grounds near Sandusky, Ohio (Figure 17-4). The project was started in 1974 with a tower 30 m high and blades 38 m in diameter. Its production capacity is 0.1 megawatts with a 8 m/sec wind or 180,000 kilowatt-hours per year. The electricity is in the form of 460 volt, 60 cycle, AC current. Some cities, such as Roswell, New Mexico, have installed such a system to help supply their electrical requirements. As energy costs continue to

Figure 17-4. A large aero-generator has been constructed by NASA at the test site near Sandusky, Ohio. (Courtesy of NASA.)

climb, such alternate sources of energy will continue to become more practical.

The available wind power at any site is related to the wind velocity; actually, the cube of the wind velocity. The midwestern United States is one of the best locations for using wind power because of greater wind speeds (Figure 17-5). A power curve can be plotted for any location that consists of the cube of wind speed plotted against the hours per year when such wind speeds were observed (Figure 17-6). Although high wind speeds are preferable, they are limited by a maximum speed when damage would occur to the propellers just as a minimum speed is required to start the propellers. Such a power duration curve shows the maximum amount of energy and the total energy that can be produced by a wind system at a particular location. It is possible to increase the available energy at any location by using a taller tower, since the wind speed increases with height.

An important application of wind driven energy systems is for individual family use. Commercially available units such as in Figure 17-7 can be purchased for a few thousand dollars. Federal and State laws may also provide substantial tax credits for their installation. Such units with 4 m diameter blades can produce 1,500 watts from 9.5 m/sec winds. Their cut-in wind speed is 4.5 m/sec and cut-out

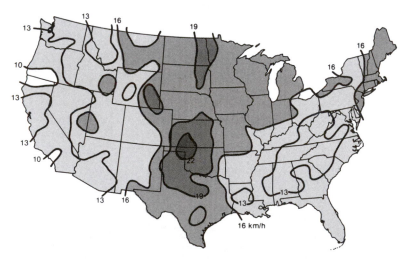

Figure 17-5. More wind power is available in certain parts of the United States than in other locations, as shown here. The central United States averages about 22 km/h wind speeds while other locations generally have lower wind speeds.

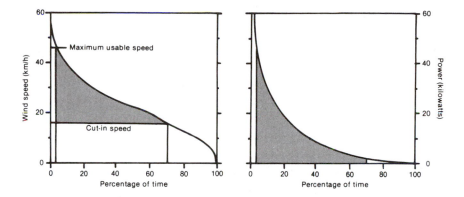

Figure 17-6. Power from the wind is related to the cut-in speed of the aero-generator and the maximum speed that the aero-generator can withstand. Average statistics can be used to compute the expected power of the wind from aero-generators.

Figure 17-7. Smaller wind systems are manufactured commercially. These may produce enough electricity to supply a major portion of the energy requirements for a single house.

speed is 18 m/sec. Although such systems can provide only about 60% of the energy for a typical home, this could represent a substantial part of the solution to our energy crisis if a large number of individuals used such units.

In mountainous terrain where the wind speeds are too high for an ordinary aerogenerator it may be possible to use a wing aerogenerator designed with no moving parts (Figure 17-8). Bernoulli's principle states that an increase in air speed is accompanied by a pressure reduction. Stationary hollow airplane wing configurations with aspirator holes can be used to obtain airflow across a turbine as shown in Figure 17-8. Such a device would operate with winds from any direction. Although icing may plug the holes and stop the generation of electricity it would not be as susceptible to damage as an aerogenerator with rotating blades. A disadvantage of this type of aerogenerator is the requirement for greater wind speeds since it has a low operating efficiency.

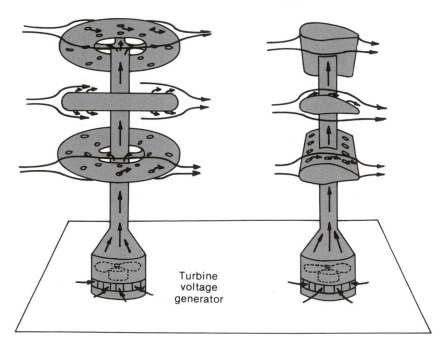

Figure 17-8. Wing aero-generators have the advantage that they have no moving parts. Their efficiency, however, is slightly less than those with blades.

Some of the disadvantages of ordinary aerogenerators are that the density of air is so low that light winds do not provide much energy; also the normal variability of winds makes this energy source discontinuous, and the capacity of turbines is limited by the size of the blades (not over 1 megawatt). A total of 100 aerogenerators each with 1 megawatt output would be necessary to equal the power potential of a small power plant.

To increase the efficiency of extracting energy from the winds, it has been suggested that tornadoes or similar vortices be utilized (Figure 17-9). A vortex has a lower central pressure than in the surrounding atmosphere and the wind speeds are much greater within the large pressure gradient. A turbine placed within the stronger winds would generate more power than one outside the vortex. According

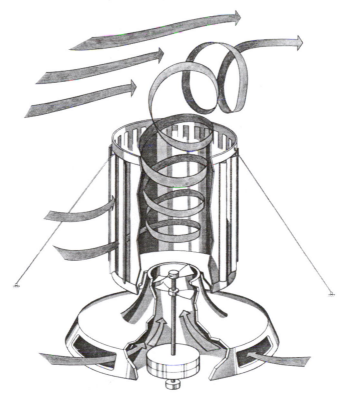

Figure 17-9. The wind can be concentrated into a vortex in order to perhaps increase the efficiency of wind systems. Reprinted with permission from *Popular Science* © 1977, Times Mirror Magazines, Inc.

to *Popular Science* (January, 1977), large cities may someday have large vortex generators, perhaps 600 m high and 200 m in diameter, nearby to convert ordinary winds to a tornadolike vortex to run a turbine. Placement of the blades that drive the generator would be very important. The Energy Research and Development Administration is sufficiently interested in this concept to have invested almost $200,000 for development to the experimental stage.

SOLAR ENERGY

In order for new energy sources to be fully utilized, they must satisfy certain environmental constraints. Solar energy, like wind energy, does not suffer from environmental limitations. The amount of available solar energy is enormous. The solar constant of 1.38 kilowatts is always available outside our atmosphere on each square meter of surface oriented perpendicular to the sun's rays. An average of half this amount gets down to the earth's surface; but even at this rate only a few square meters of absorbing area can collect enough energy to provide all the needs of a typical American home. A house with a peak load requirement of 2.5 kilowatts would need only 3.6 m^2 of collectors if their efficiency were 100%.

It is estimated that the United States will need 2,500 megawatts of power by the year 2000. Assuming 50% of the sunlight penetrates the atmosphere to fall on collectors that are 50% efficient only 8 km^2 would supply enough energy for the entire United States.

Areas most suited for solar heating are those with little cloudy weather and plenty of sunshine. Figure 17-10 shows the potential for solar heating in January based on the relative amount of heating that could be supplied to a building. The lowest potential for solar heating exists near Lake Ontario with much greater potential throughout the southern half of the United States.

Three basic methods are used to harness solar power. One is the process of photosynthesis that plants use in the presence of sunlight to convert water and air into plant tissues. Others are solar collectors and solar cells. Since the process of photosynthesis cannot be duplicated in the laboratory the remaining methods are utilized.

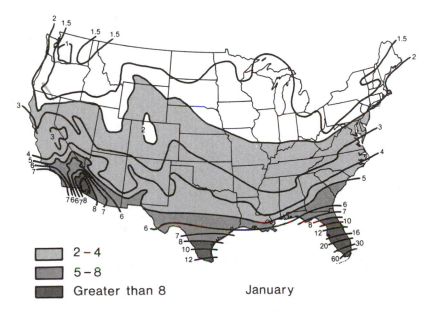

Figure 17-10. The relative potential of solar heating in different parts of the United States is shown by the numbers here. A rating of 4 means that a given solar collector in that region could supply 4 times as much of the needed energy as it could in the Ontario region, for example. (After Hoecker, NOAA.)

SOLAR CELLS

The technology exists for producing functioning solar cells. The power source of Skylab was an array of solar cells designed to provide 20 kilowatts of power (Figure 17-11).

Solar cells are made from silicon, the second most abundant element in the earth's crust. Pure silicon crystals separate electrical charges when placed in sunlight as a result of the photoelectric effect first explained by Einstein. Photons within sunlight release electrons from the silicon atoms. These electrons can be made to flow through a wire to provide an electrical current. Since the silicon is not depleted the process can be repeated continuously. One of the problems associated with them is the cost. Since solar cells large enough to produce one kilowatt of electricity cost about $4,000, it would

Figure 17-11. The solar cell power supply used by Skylab was one of the largest ever produced. Solar cells ran the space station in spite of the loss of one of the panels during launch. (NASA Photograph.)

require $10,000 to equip a house with solar cells to provide 2.5 kilowatts of electricity.

Another problem is a means of storing energy for days when the sun is obscured by clouds. The most common storage system is a battery of the lead acid type. One kilowatt batteries may become commercially available in the 1980s, but it may be a cube 2 m high. More than one battery of this size would be required to store a day's supply of energy. Research is being directed toward the development of new lighter weight batteries using materials that are not so corrosive.

SOLAR COLLECTORS

Solar energy can also be obtained in a usable form from collectors. A flat plate collector consists of a black surface for absorbing sunlight behind one or more glass cover plates (Figure 17-12). Air, water, or other fluids are circulated between the absorbing plate and the cover to extract the heat.

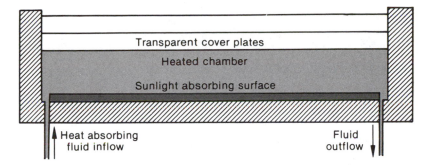

Figure 17-12. The components of a solar collector are shown here.

The handyman can construct his own solar collector by attaching empty soda cans to a sheet of plywood, painting this a dull black, and covering it with a sheet of plate glass positioned at least a few centimeters above the metal cans. The edges can be closed except for an intake and outflow hose. If a fan is connected to circulate air between the plates and into a room it becomes an active solar system. If the warm air must move by itself through a duct leading from the top of the plates it is a passive system.

Solar collectors should be oriented at the proper angle to take advantage of the most direct sun's rays. This depends on the latitude of your location. At 40° latitude, the sun's angle is 74° from the southern horizon at its highest point on June 21 at noon. It is only 27° up from the southern horizon at noon on December 21. A stationary solar collector used primarily for heating in winter could be turned to the south so that it makes a 90° angle with the sun's rays or about 60° with a horizontal surface.

For optimum absorption of sunlight a collector can be mounted to rotate with the sun. Correct orientation can be provided by hand or by computerized systems that are already commercially available.

Measurements have shown that temperatures greater than the boiling point of water may be achieved within a flat plate collector with outside temperatures below freezing. Therefore, circulation of air or water through the collector may be necessary to prevent excessive heat buildup and possible damage to materials used in its construction.

A common solar heating system consists of flat plate collectors with water circulation for heating the house and providing hot water. If a large storage tank is used it has the added advantage of heat carry-over for cloudy days. An indoor swimming pool can also be heated as a means of retaining heat. This method of solar heating is largely a plumbing problem, with antifreeze required if the temperature of the water approaches the freezing point. It is necessary to have an auxiliary heat supply for extended periods of cloudy weather at many locations.

An alternative heat carrier is air. Although it is not as efficient because the heat storage capacity of air is much less than water, it has the advantage of causing less trouble if it leaks out of the system. Air may be circulated directly through the collectors and redistributed throughout the house by ducts. Heat storage is required since you would not want the heat coming through the house on a warm day. Half the basement may be filled with rocks for the air to circulate through to retain heat until it is needed. Thus, the air would carry heat from collectors to the storage area on warm days and from the storage area to heat the house on cold days (Figure 17-13).

A similar idea can be used with water as the heat carrying medium. A solar collector built on the south side of a house could utilize

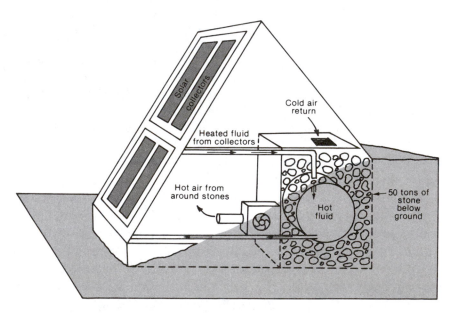

Figure 17-13. A solar house may utilize solar collectors to heat a fluid. This fluid is then circulated through a reservoir and the extracted heat is distributed throughout the house.

ordinary galvanized metal roofing with ridges as the metal to be painted black to collect heat. Water pumped to the top to trickle down the corrugations would extract this heat. A plate glass placed over the collector would prevent heat loss. The hot water should be circulated to a storage area in the basement composed of fifty tons of rock around a large tank. Hot water from the collectors is circulated through the tank during the winter. A second circulation system sends air through the heated rocks to be distributed throughout the house. Such a system can provide heating during several days of cloudy cold weather. It can also be used for cooling during the summer as well. If the water is circulated through the collectors at night when cooling is provided by the cool metal it would cool the storage tank of water and surrounding rocks. Then, during the hot day, air flowing through the storage area would be cooled as a source of cool air to be distributed throughout the house.

It is possible to use solar heating in other ways such as in solar furnaces. These can be constructed by using a number of dish shaped mirrors to concentrate sunlight on a boiler. Water can be heated to the boiling point for the generation of steam. This is then a steam engine fired by the sun instead of coal or other fuel. Steam can, of course, be used to create electricity, circulate through radiators to heat a house or for other purposes.

SOLAR ENERGY ACCEPTANCE

The conversion to solar energy in homes and industry is very slow, more because of economical and sociocultural than technological factors. The initial cost of a solar system for a home is several thousand dollars more than a conventional system. This factor, along with the inability of a solar heating system to provide the entire heating during the coldest and cloudiest months throughout most of the United States, accounts for much of its slow acceptance rate. In addition, a solar collector is large enough to be noticed and most people do not want an appendage on their house unless it is very attractively packaged. A solar system may be designed along with a house prior to construction to produce a modern integrated design (Figure 17-14). This may improve acceptance of this alternate form of energy in the future.

Figure 17-14. Solar collectors can be designed into a house to form an integral part of it. This house was built by Kansas Power and Light Company for testing purposes.

Economics of solar systems are related to the price of conventional energy and the demand for solar units. Both these factors are changing with time in a direction favorable for much greater utilization of solar energy either in the form of collectors or solar cells.

SUMMARY

Since our energy supply will probably continue to be short, other sources of energy may become increasingly important. A tremendous amount of energy is involved in many atmospheric processes. A single thunderstorm contains an enormous supply of energy, but practical means of harvesting this energy have not been developed. However, wind systems are available commercially and may be important in supplying a significant proportion of our future energy needs.

Large experimental aerogenerators are now in operation. One of these being tested by NASA has a production capacity of 0.1 megawatts. Such large wind systems could supply a major part of the electrical requirements of a city. Smaller systems, with perhaps 4 m diameter blades, are being placed in use by individual families. These systems may provide more than half the energy requirement for a typical home. Research is being conducted on improving the efficiency of wind systems, as well as on new approaches. A turbine located within a large cylinder used to create a vortex, for example, might be expected to improve the amount of energy extracted from the wind.

Solar energy is another important alternate energy source. The amount of energy available from the sun is enormous. Solar cells may be used to extract the energy from sunlight. These may be constructed of silicon crystals that generate electricity as a result of the photoelectric effect. Since this process does not deplete the silicon itself, it can be repeated indefinitely. The major current problem with solar cells, however, is their cost.

Solar collectors are another means of converting sunlight to a useable form of energy. Solar collectors consist of an absorbing surface with air, water, or other fluid circulated between the absorbing plates to extract the heat. Thus, a flat plate collector may be mounted on the roof of a house with air circulated through it to remove the heat and carry it into the house for heating the rooms or warming water. The storage of heat for cloudy days is an obstacle for either photocells or solar collectors.

Various methods such as the storage of electricity from photocells in batteries, or the storage of heat from solar collectors in tanks filled with water have been used to operate solar systems on cloudy days. As these techniques and others are perfected, solar energy may become an important factor in supplying energy.

18
Acid Rain, Ozone Hole
and Greenhouse Effect

It has been so hot for September. I wonder if the greenhouse effect could be causing this unusual weather?

While I was outside last week I came in with a severe sunburn after spending too much time looking at the south wall of our limestone house. It appears to be eroding faster than the other walls. Could this be due to acid rain? And could my sunburn be caused by a weakened ozone layer?

OCCURRENCE OF ACID RAIN

Pollutants which are injected into the atmosphere, such as the SO_2, NO_2 and NO gases, are transported great distances downwind where they are transformed into sulfate and nitrate ions which are tiny particles. They enter into the precipitation process and also undergo dry deposition. The result is a large number of effects, such as the increased acidity of impounded water. Measurements indicate that the acidity is rising in many lakes. In Scandinavia 10,000 lakes were reported to be devoid of fish and this was attributed to the SO_2 from England, France, and Germany. In the United States the pH fell from 6.5 to 4.8 over the last 40 years in the Adirondack Mountains in New York. It is estimated that 200 lakes are dead in the high latitude regions in the state of New York because of acid rain which is falling over the area. Some forests are affected through dry deposition and through changes in the pH of the rainwater that falls on the soil.

Buildings and monuments deteriorate much faster with acid rain because of the combination of wet and dry deposition of pollutants. Some historically important structures in Europe are significantly affected. Porous building materials and limestone are damaged most as acid rain and particles are held in the porous material. In addition to limestone, other materials including concrete, iron, steel, zinc and paint are all affected by acid rain. There is some indication that the acid water in lakes produces an elevated concentration of toxic metals that may enter into the food chain and therefore affect people.

Acid rain is defined as rain which is less than 5.7 pH. The acidity of rain water varies considerably for different locations. In regions where high alkaline dusts are present, it may be neutral, about 7, but the pH is typically around 6. In comparison with other liquids, vinegar has a pH of about 2 and battery acid has a pH of 1. On the basic end of the scale, baking soda has a pH of slightly more than 8 and ammonia has a pH of 11.

CAUSE AND DISTRIBUTION

Many measurements have been made of the acidity of rainfall over various parts of the United States. The distribution of acid rain is shown in Figure 18-1. A region of very low pH values for rainfall extends over the northeastern United States. The average rainfall in many of the New England states has a pH less than 4.2 and all of these states have rainfall with a pH less 4.6. Rainfall over the rest of the United States has typical values of about 5 or slightly over 5. The large geographical differences in the acidity of rain water extend into Canada, where parts of Canada located near the Atlantic Ocean have experienced increases in the acidity of rain water.

The cause of acid rain is primarily the injection of material into the atmosphere by human activity. Sulfur and nitrogen compounds from the burning of fossil fuels may be deposited on the ground or on forests. Sulfate and nitrate ions may fall as dry deposition where they affect vegetation, buildings, and materials directly, or they may be incorporated into the rain formation process and then fall with the rain. In general, acid rain refers to both processes; dry deposition as well as the lowered acidity of rain water.

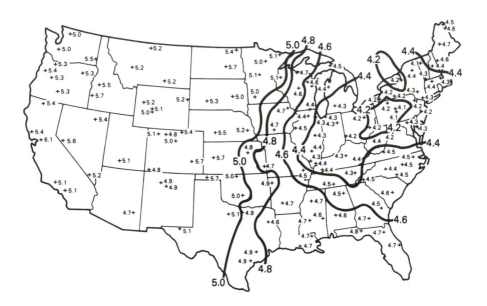

Figure 18-1. The distribution of acid rain in the United States (From Annual Report, National Acid Precipitation Assessment Program, *Washington, D.C., 1989)*

Primary sources of sulfate and nitrate ions in the atmosphere are electrical power generation, industrial processes, automobiles, and other human activity. The contribution from natural sources includes emission of these ions from swamps, volcanoes and the ocean. These sources are generally quite small in comparison to the number of sulfate ions emitted into the atmosphere from human activity.

The precursors of the sulfate ion are primarily SO_2 and hydrogen sulfide gases. These gases are transformed through various chemical reactions into particles which precipitate out as sulfate ions. These small particles (SO_4 with two negative charges) then either enter into the precipitation process or fall as dry deposition on the earth's surface. Natural sources for the nitrate ion (NO_3 with one negative charge) are lightning and biological activity in soils. Precursors for the nitrate ion are the gases NO and NO_2. These gases are transformed into small solid particles which again undergo dry deposition or enter into the rainfall formation process.

Other ions may also contribute to acid rain. Most of these are insignificant in comparison to the sulfate and nitration ions, but include chloride, ammonia, sodium, calcium, magnesium, and potassium. Chloride and sodium ions may arise from the common practice of putting salt on streets, for example, and a small number of these find their way into the atmosphere. Oceans may also contribute some sodium. Agricultural and industrial processes may contribute magnesium and potassium to the atmosphere, but the influence of all other ions on acid rain is quite small in comparison to the sulfate and nitrate ions.

The resulting acidity of lakes, streams, and ponds which accumulate acid rainwater is greatly influenced by the amount of calcium in the surrounding rock formations. Such calcium gives some lakes and ponds a large buffering capacity. The weathering process releases some calcium from limestone rocks and the rain that falls on calcium rich soil is neutralized by calcium incorporated into the runoff as it flows into streams. Similarly the river beds and lake beds formed from limestone rocks allow the water to dissolve some of the calcium and this neutralizes the acid rain. Therefore, a lake that is surrounded by limestone has a large buffering capacity for acid rain.

In general, the Northeastern United States has very little buffering capacity, because of the strong weathering of rocks over many years. Therefore, this region is more prone to the problems of acid rain than other locations, not only because of the large number of power plants that are putting sulfur and nitrogen gases into the atmosphere, but also because of the low buffering capacity of the region.

POLLUTION AND PRECIPITATION PROCESSES

Air pollution is likely to have a variety of effects on the raindrop formation process in both warm and cold clouds. For warm clouds the first pollution effect is that of providing more condensation nuclei. The second effect is a greater growth rate of the more concentrated tiny droplets due to the increased solute effect. Increased pollution also changes the drop size distribution and therefore, affects the coalescence process, whereby cloud droplets grow into raindrops. The polluted atmosphere may also provide more sulfate and nitrate ions which serve as condensation nuclei and produce acid rain.

In cold clouds pollutants may add additional ice nuclei or they may add salts which lower the vapor pressure of liquid droplets and reduce the growth of ice crystals. As the ice crystals grow to snowflakes, increased pollution may cause increased clumping, and increased melting. When acid pollutants serve as the ice nuclei, cold clouds also produce acid rain or acid snow at the surface.

When the entire precipitation process is considered it seems amazing that precipitation amounts are not changed more by pollution than observations would indicate. Numerous investigations have been conducted on the amount of rainfall in polluted areas compared to unpolluted areas. The amounts do not seem to be altered enough to be detected even in more polluted areas.

POLITICS OF ACID RAIN

The politics of acid rain are interesting. The Canadians and the United States agree that about 50% of the acid rain falling in Eastern Canada originates from United States sources. At the same time about 25% of the acid rain in the New England area of the United States originates from Canadian sources. So the transport of pollutants across the boundaries of these countries becomes a political issue. Conferences have been held on the subject and negotiations continue concerning the source of pollutants, and solutions to the acid rain problem. Canada has agreed to cut in half its relevant air pollutants by the year 1994 and has requested that the United States do the same. In the United States it is not a simple matter to reduce these pollutants by such an amount.

One solution would be to install wet scrubbers in all the stacks that emit SO_2. The installation of scrubbers is very expensive; it has been estimated that about $20 million is required to install a scrubber in an existing power plant.

Another solution would be to convert power generation plants to nuclear power. A study by the World Watch Institute showed that a new nuclear power plant can generate electricity for about

11¢ per kwhr where a new coal burning power plant could generate it for about 6¢ per kwhr. If existing power plants that burn fossil fuels are required to install expensive equipment then this could eliminate the difference in cost of electricity produced by coal burning and nuclear power plants.

One estimate indicates that an increase in the electric bills of all U.S. citizens of only 2% would be enough to take care of the acid rain problem. Another estimate indicated that if the controls were concentrated in the Midwestern and Eastern United States, where the sources of pollutants are causing the acid rain problem, the cost would be a 30% rise in electricity rates. Such an increase is so great that it is not likely to be tolerated without considerable resistance.

OZONE HOLE

Chemicals of human origin, particularly chlorofluorocarbons (CFC), have received considerable public attention because of the threat to the ozone layer. These gases are used in aerosol spray cans for deodorants and hair spray in several countries, although their use has been banned in the United States. The CFC gases rise into the stratosphere where they lead to photochemical reactions that destroy the naturally occurring ozone that protects us from the harmful effects of solar ultraviolet radiation which include sunburn and skin cancer.

In 1985 British Scientists discovered the Antarctic ozone hole which has lead to increased concern about the ozone layer. Much of the ozone above Antarctica disappears each austral spring (Figure 18-2) and this decrease has been linked to CFC.

NOAA scientist have developed a program for continuous global monitoring of ozone from polar orbiting satellites. Since 1985 measurements have been made for the daytime total and vertical distribution of ozone with a precise Solar Backscatter Ultraviolet instrument. This instrument measures the spectrum of solar ultraviolet radiation scattered back to space by the earth's atmosphere, which depends on the amount of radiation absorbed by the ozone layer.

In addition to satellite measurements of ozone, NOAA has received and archived total ozone amount and vertical profiles

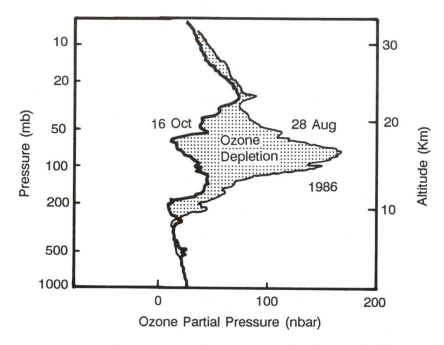

Figure 18-2. The amount of ozone over the Antarctic decreases each spring (October) establishing the "ozone hole" (after Hofmann, 1987).

since 1985 from the global network of Dobson spectrometer and balloonsonde measurements. Comparisons of total ozone on a monthly basis show general overall agreement between different methods of measurement but with some time dependent differences.

Several theories have been advanced for the formation of the ozone hole over Antarctica. It has been related to the presence of CFC, the circumpolar vortex and polar stratospheric clouds. A strong circumpolar vortex around the South Pole prevents much of the north-south mixing of air. The breakup of the circumpolar vortex in winter is associated with more ozone in Antarctica. The ice clouds in the polar stratosphere affect the ozone layer through chemical reactions involving the ice crystals and by shielding the atmosphere from sunlight which affects the concentration of ozone.

Perhaps the release of CFC into the atmosphere will be curtailed before concentrations are great enough to affect the ozone layer of the rest of the world. The Arctic ozone layer would be expected to be affected before midlatitude and tropical regions of the world. There is little evidence to indicate that these regions are currently affected in the same manner as Antarctica. One of the reasons the Arctic ozone layer is not affected as much as its southern counterpart is that the circumpolar vortex is not as well developed there because of more land and greater roughness of the surface.

THE GREENHOUSE EFFECT

The greenhouse effect is caused by the absorption of earth radiation by gases in the atmosphere while these gases remain transparent to sunlight. The concentration of CO_2 in the atmosphere is responsible for much of the greenhouse effect. Water vapor is also a very important absorber of infrared radiation from the earth, but it generally has been ignored in comparison to CO_2, perhaps because of its continuous evaporation and condensation.

Measurements of CO_2 levels in the atmosphere have been made for several decades at the Mauna Loa Observatory in Hawaii. It is estimated that the CO_2 levels in the atmosphere were approximately 285 ppm at the beginning of the Industrial Revolution, about 1890. Measurements at Mauna Loa show that the CO_2 levels have increased from 315 ppm in 1958 to almost 340 ppm. This is attributed by most people to the addition of 5 billion metric tons of carbon per year to the atmosphere through the burning of fossil fuel.

CONCERN FOR GREENHOUSE WARMING

In 1983 the Environmental Protection Agency released a report called "Can We Delay a Greenhouse Warming". The very title of this report assumes that a warming is imminent and we must immediately do something to prevent that warming. The report cites a previous National Academy of Sciences report in 1979 which concluded that a doubling of CO_2 levels would increase global air temperatures by 3°C. The Academy reconvened in 1982 and found no reason to change the conclusion from the first report. Therefore, the EPA report starts with a premise of $3°C$ warming for a doubling of CO_2 levels and points out some of the implications of this warming.

One of the effects of global warming would be a rise in sea level. It is estimated that an increase of 2 to 11 ft will occur in the next 120 years. It would also be expected that changes in precipitation would occur, that storm patterns would be altered and because of the increase in temperature, the growing season of crops would be changed. Some beneficial effects of increasing CO_2 levels could be expected. Photosynthesis of plants would be enhanced due to the greater quantities of CO_2 in the atmosphere. The climate of higher latitudes would be improved due to warming. There would be a reduction in the heating cost globally because of increased temperatures.

GREENHOUSE GASSES

Although CO_2 is generally considered to be the primary greenhouse gas, other greenhouse gases besides CO_2 and water vapor may also be important. These are nitrous oxide, methane and chlorofluorocarbons (CFC). Nitrous oxide is a product of biological denitrification which occurs in the soil and oceans. The production of methane also occurs naturally due to anaerobic fermentation in such locations as rice fields and swamps. The input of CFC, however, is almost entirely a product of human activity, with 58% of the emissions coming from the propellant used in deodorants and hair sprays.

FUTURE ATMOSPHERIC TEMPERATURE

The baseline temperature estimates in the EPA report were made by assuming a $3°C$ increase in temperature for a doubling of CO_2. For the baseline estimate, a $2°C$ increase in temperature would occur in the year 2040. If the atmosphere were less sensitive to CO_2 levels and warmed only $1.5°C$ for a doubling of CO_2 levels, the time would be increased to 2070 for a $2°C$ warming. If on the other hand, it is more sensitive to CO_2 and the atmosphere increased $4.5°C$ for a doubling of CO_2 levels, and with a moderate increase in other greenhouse gases, the $2°C$ increase in temperature could occur as early as 2030.

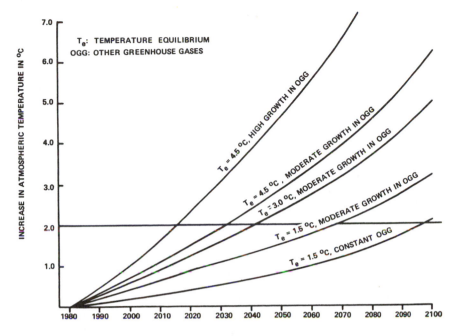

Figure 18-3. The increase in temperature in future years esti-mated for various scenarios. (From Seidel, S. and D. Keyes, Can We Prevent a Greenhouse Warming, EPA, 1983)

The EPA report also calculated the changes in temperature projected to the year 2100 and considered various alternatives that would cause the temperature warming to deviate from the mid-range baseline temperature of 5°C (Figure 18-3). Greatest effects would be produced by a high growth of other greenhouse gases which would produce a warming of 8.3°C. In contrast, if the temperature is affected less by the doubling of CO_2 by an amount of 1.5°C instead of 3°C, warming would be only 2°C by the year 2100 instead of 5°C. Similarly a ban on coal and shale oil would reduce the 5°C projected temperature warming by the year 2100.

OMITTED FACTORS IN THE EPA REPORT

Many questions remain that aren't addressed in the EPA report. One of these is whether the amount of cloud cover is chang-ing enough to counterbalance the greenhouse effect, since cloud

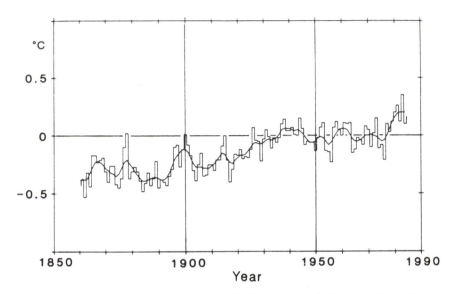

Figure 18-4. Temperature trends for the whole earth. (From Jones, P.D., T.M.L. Wigley and P.B. Wright, "Global Temperature Variations Between 1861 and 1984," Nature, 1986.)

cover will reduce solar radiation. Another question is whether small particles are increasing enough to offset the CO_2 effects. Another question concerns the fact that we have had significant increases in CO_2 levels since the Industrial Revolution, and this should have already produced measurable temperature changes. An article by Jones *et al.* considers the temperature for the world from 1861 to 1984 and gives the temperature trends for the whole world (Figure 18-4), the northern hemisphere and the southern hemisphere. These show a gradual increase in temperature to about 1940, and then a decrease in the global temperature from 1940 to the mid-1970's, and an increase from the 1970's until 1984. This variation does not match the expected changes from CO_2 increases very well.

Another consideration is our current position with respect to glacial and inter-glacial periods. Geologists tell us that we are in the warmest time period ever experienced in history and according to past long-term cycles should be heading toward much colder temperatures. Therefore, one possibility is that the greenhouse warming will offset the temperature slide that we would have expected due to long-term cycles.

Jones tried to address another of the important questions related to the effect of the urban heat island on temperature data obtained at the same location over a long time period. The urban heat island is too large to ignore. For example, Kansas City has an average of 4°C increase in temperature in the downtown area compared to surrounding rural areas; Topeka has an increase of 3°C and Lawrence, Kansas, has a temperature increase of 2°C. The temperature increase has been shown to be related to city size. The length of time required for a city to grow from 50,000 to 250,000, which would represent a 1°C increase in temperature may be only a few decades. Therefore, the increasing urban heat island as the city grows must be considered.

Jones considered the urban effect only by omitting 38 stations out of 2,666 in the Northern Hemisphere where he thought the growth of the city was a problem. It is unlikely that this was sufficient to take care of the effect of the urban heat island. If the change in global temperature of 2°C which some investigators expect to occur by the year 2040 is to be measured, the larger effect of the urban heat island must be considered for every city which has grown in size during the period of temperature measurements, and this includes most cities.

Another investigator, Idso, has suggested that the increases in CO_2 levels are producing a cooling effect instead of atmospheric warming. He used the reasoning that the greatest increases of CO_2 have occurred since 1940 and the corresponding trend in temperature since 1940 would indicate that increasing levels of CO_2 are resulting in atmospheric cooling rather than atmospheric heating. The physical mechanism could be related to evaporation rates and cloud cover formation.

Other research since the EPA report in 1983 has dealt with the changes that would be produced by the increased atmospheric temperatures caused by the greenhouse effect. For example, an article by Washington and Meehl used the NCAR Community Climate Model to look at the effects of a doubling of CO_2 levels on snow and sea ice and their effect on albedo (reflection of sunlight) since this would change the global average temperature. This model calculates that an increase in air temperature would occur with doubling of CO_2, but the type of snow and sea ice formations would affect reflectivity and thus, the amount of warming. They also found that colder initial conditions produced a greater warming due to increasing CO_2.

In another article (Henderson-Sellers) it was found from observations that cloud cover has generally increased during both a cold period (1901-1920) and a warm period in Europe (1934-1953). They report that climatic simulation models have given no agreement for either direction or amount of change in cloud cover for a "warming world."

It is apparent that further research must be completed before all the answers are available concerning the greenhouse effect. Before we can trust computer models to provide the answers they must be able to correctly model all important aspects of the problem. There is no evidence that current models have been able to accomplish this. Notable missing factors are the temperature effects of changing cloud cover and increasing small particulates in the atmosphere.

SUMMARY

Some scientists believe the weather will become increasingly "unusual" in the future because of the burning of fossil fuels and the use of chlorofluorocarbons in spray cans. Specific problems caused by pollution of the atmosphere are acid rain, the ozone hole and an enhanced greenhouse effect.

Acid rain is a problem which threatens forests and lakes in the Eastern United States and other parts of the world. It also causes limestone buildings and monuments to deteriorate more rapidly. Its primary cause is the injection of sulfur and nitrogen gases into the atmosphere from the burning of fossil fuels such as coal and oil.

The injection of chlorofluorocarbons into the atmosphere, primarily from deodorant and hair spray cans manufactured outside the United States, leads to photochemical reactions in the stratosphere that destroy the naturally occurring ozone. The formation of the "ozone hole" over Antarctica during the spring months has been linked to the presence of CFC, the circumpolar vortex and polar stratospheric clouds.

The greenhouse effect is enhanced by increased amounts of CO_2 in the atmosphere. If the increasing CO_2 levels were the only factor affecting our climate the calculated temperature increase of 2°C would occur in the year 2040. However, the earth's climate is determined by a large number of interdependent factors. The combination of such factors as CO_2 levels, cloud cover, albedo and particulate levels in the atmosphere will determine whether our future climate will be different or remain unchanged by human activity.

19
Human Response to Weather

Now that the season's here again; it seems worse every year, when every advancing squall line fills everyone with fear. The weather station radar shows a certain kind of blotch, and the whole surrounding area goes on tornado watch. One hundred miles on either side of a line from here to there marks out the ill-starred region to be ravaged from the air. The picture tubes and radios spew forth the voice of doom; uneasy folks peer nervously through the deepening gloom. Children sense their parents' fear, begin to cry and fret; each ominous announcement reamplifies the threat. To a terror stricken observer with myopic sight endowed a trailing wisp of vapor looks just like a funnel cloud. He quickly calls the station, relating what he has seen, and the "citizen observer" has brought panic to the scene! Folks dash to the basement, or cower 'neath the bed as the word "tornado warning" fills anxious hearts with dread. Our storm watch broadcast network saves many lives, 'tis said, but the frightened die a thousand deaths, before they're really dead. Of the thousand ways of dying, if you analyze them all, the chance of being blown to death is ridiculously small. So when the voice assails me, "a fearful storm is brewing," I remember the statistics, and continue what I'm doing. If the awesome funnel gets me, as in the end it might, I'll have lived my life serenely, not quivering in fright.

—Charles Lacey, Lawrence, Kansas

A very common response of many people to severe and unusual weather is to ignore it and assume that it will never affect them directly. This assumption may cause a person to exhibit such unconcerned behavior as to continue taking photographs of a tornado, prior to, and after the tornado literally blows him off his feet, as happened during the Wichita Falls Tornado. It can also cause a person to spend his Thanksgiving vacation in a community building in a strange city while a blizzard rages outside as happened to hundreds of people in 1979 as a blizzard produced drifts several meters high to close highways and strand thousands of people.

BLIZZARDS

Many people do not have an appropriate respect for the potentially lethal power of the blizzard. Thus the blizzard is underestimated in contrast to the tornado that is frequently overestimated in terms of its lethal ability. In fact, both of these types of storms cause about the same number of deaths each year. A reasonable response to the threat of blizzards is to recognize their capability of closing major highways, and making travel on streets impossible. Thus, one who insists on traveling through a blizzard as it moves across the United States must realize that it has the capacity for reducing the heated interior of his automobile to outside temperatures by stalling the car where it will run out of gas in a few hours if it is used to keep the heater going.

On November 21, 1979, a blizzard cyclone was dumping over one-half meter of snow in Cheyenne, Wyoming and Denver, Colorado, before starting to progress eastward. It continued to create blizzard conditions with low temperatures and heavy snowfall. It was no secret that this winter storm would be moving across the Central Plains at a time when many people would be traveling for their Thanksgiving vacations. The inevitable result was stranded motorists and 125 people killed. National Guardsmen in Wyoming rescued more than 2000 travelers – a stalled caravan 4 km long – on Interstate 25 north of Cheyenne. Several had to be rescued by helicopter. Motels and hotels became jammed throughout the Midwest as the blizzard moved through Kansas, Nebraska, Iowa, and Minnesota. Many people spent their Thanksgiving stranded in their cars, in motels, National Guard Armories, or municipal auditoriums. In Sidney, Nebraska, about 60 motorists were treated to a turkey dinner by a local church group. Many others were lucky just to keep warm.

Only a little more than a year earlier on January 26, 1978, another blizzard had moved through the Midwest and killed almost as many people. More than 5,700 motorists were stranded along snow-blocked highways in Ohio alone, as Governor James Rhodes called the storm "a killer blizzard looking for victims." The freezing temperatures were an added difficulty to many other Ohio residents as more than 150,000 homes were without electricity for heating for most of the day on January 26th. A federal state of emergency was declared by President Carter and the 5th Army moved in to help stranded motorists and exhausted utility repairmen.

Such examples occur more frequently than most people realize and emphasize the need for proper respect for these winter storms. It is apparent that the proper response is to consider weather forecasts when making plans to travel during winter, and to plan around major winter storms.

TORNADOES

The specific response of a person to tornadoes, in general, is as varied as the number of different types of personalities. Some people who live in cities that have been struck by tornadoes, and other people who are only vaguely aware of tornadoes, are terrified by them. Others consider them a thing of beauty to be included in a work of art or to be photographed as one of nature's most beautiful creations. Most people are more afraid of tornadoes than is warranted by statistics on deaths and injuries caused by them. This, no doubt, is because of the very violent nature of tornadoes.

Many factual stories are circulated after a tornado that may reinforce or initiate fear of these storms. Less publicized, however, is the fact that a typical severe tornado can destroy one thousand houses, most of which have occupants in them, and kill only twenty people. If the destroyed houses contained an average of 3 persons per house, then only 20 people out of the 3000 potential victims would be killed. Based on this example, your chances of being killed by a tornado are less than 1% even if your house is totally destroyed by a tornado.

A particular person's chances for survival during a direct hit by a tornado are certainly improved by applying the best information on safest locations in houses and by properly designing houses and basements for shelter from storms. Statistics show that these responses combined with the increasing accuracy of warnings and forecast information are resulting in fewer deaths each year from tornadoes even though property damage from tornadoes continues to increase.

LIGHTNING

Many people's response to lightning consists of placing a lightning rod on their house or office building as protection from a lightning strike or obtaining insurance for personal property. Many people do not realize that lightning is responsible for as many or more deaths each year than tornadoes; since these deaths are usually single events they do not make the headlines as often as many other storm related deaths. Individual responses to lightning are probably more rational than reactions to many of the other potentially fatal forms of severe weather. This is probably because the threat from lightning can be recognized by the loud noises and visual displays that accompany thunderstorms. Therefore, if people are aware of the most effective precautions during lightning displays they are more likely to put them into practice.

HAILSTORMS

Major human responses to hailstorms have consisted primarily of insuring personal property or crops, and seeking to modify hail producing thunderstorms. The threat of damage from hailstorms varies with geographical location but in many areas a significant cost of crop production is insurance to pay for hail damages, should they occur. Automobile insurance and house insurance normally provide coverage from hail and wind damage in most parts of the United States.

Even though the National Hail Experiment concerned with modifying hail producing thunderstorms was terminated prematurely because of lack of positive results, weather modification for the purpose of reducing hail damage is still practiced by some agricultural groups in the United States. This is likely to continue because of the large potential benefit.

HURRICANES

Our ability to respond to the threat of hurricanes has been greatly improved in the United States since the age of satellites. Since satellites are able to photograph cloud patterns several times each hour for early identification, their long life allows greater preparation time and a more rational response.

Forecasting the exact location of landfall is still a limitation for specific warnings since every effort is made to prevent unnecessary evacuations of cities. The number of people that are forced to evacuate low-lying coastal areas for a typical strong hurricane is about a quarter of a million people. When this number of people are involved, it is not surprising that a few brave but unenlightened people decide to remain behind for a hurricane party as a large storm approaches.

With today's mobile society, it is estimated that more than 75 percent of the population along coastal areas of the United States are inexperienced with hurricanes. In addition, a person who has experienced a previous hurricane is conditioned by that experience. This is not always helpful since many weak hurricanes strike the United States and give a person riding them out a wrong impression of the possible intensity of future storms. Therefore, it is important to realize that hurricanes come with a variety of intensities. When a strong hurricane approaches land, the only rational behavior is evacuation of low-lying areas when advised to do so.

Methods of evacuation and evacuation routes therefore become important. Methods of evacuation are a major limitation in many less developed and overpopulated countries. In the United States, however, the automobile offers a rapid means of escaping the fury of these storms. Evacuation procedures and routes have been prepared for many counties along the coast to prevent problems that can develop from short warnings and overcrowded highways. Individuals may also seek to protect property by boarding up windows, or by purchasing metal hurricane protectors that are available commercially to help cover buildings during a hurricane.

DROUGHT

Response strategies to drought are more difficult to plan than for some of the other forms of severe weather. Irrigation systems are frequently used to eliminate drought. However, such solutions require a source of water and also considerable lead time for putting the system into operation.

In the United States, government assistance is frequently provided to drought stricken areas, just as it is for victims of other forms of severe and unusual weather. Within the semiarid parts of the United States and several other countries, such as Africa, it is not unusual to find substantial efforts in weather modification. This represents group activity aimed at broad solutions to the drought problem.

Nations may respond to the threat of drought by the storage of grains and other food products for use in the event of an extended national drought. City governments may react to a drought problem by first requiring the conservation of water, followed by rationing of the water supply. Continued drought may fan the flames of proposals such as piping water in, desalination of sea water, or dragging down an iceberg.

Within the agricultural community, those affected by drought may decide to alter their cropping strategies or seek a temporary job off the farm. Ranchers may liquidate their herds to be replenished later, probably for higher prices, as their pastures are burned into the ground. More drought resistant crops or crops that mature faster may be additional options. Many farm ponds were dug after the dry years of the 1930s. These served as an important source of water during droughts in the 50s and also during more recent droughts.

FLOODS

A major human response to floods, as in other types of severe weather, has been to ignore the potential threat. A significant part of many cities is located within the floodplain of a large river. Housing developments and industries frequently expand into low-lying floodplains. Such planning and construction completely ignore the fact that the floodplain surrounding a river is flat because it is occasionally inundated by water. The return period for floods may be several times per decade or only once per century. As a community with developments in a floodplain realizes the potential threat, levees may be built along the river banks. As these continue to require additional height they may eventually reach the stage where the bottom of the river channel is actually above the lowlands outside the levee area.

This makes the threat of flooding even more pronounced. Dikes may also extend out into the current of a river, such as the Missouri River, to prevent it from eroding the banks. The Mississippi River Commission, established in 1879, has constructed an entire system of levees along the banks of the river. These now total more than 4000 km in length and are more than 10 m high in some places.

Response to the threat of floods on a national level has also included the establishment of flashflood warning procedures with personnel to operate river forecast offices. An additional Federal response is the construction of dams to form lakes with the intent of flood control of particular rivers. By controlling the flow of water over a dam, flood losses may be reduced or prevented downstream. If the flow of water over the dam is of sufficient quantity and height to produce electricity this is an additional benefit.

Individuals may respond to the threat of flood by recognizing that floodplains suffer from the threat of floods.   Thus, a floodplain would not be selected for a housing development if another choice were available.

Proper reaction to the threat of flashfloods is also very important. A tabulation of the 25 casualties of the Kansas City Plaza flash flood of September 12 and 13, 1977, indicated that 17 of the victims were driving or passengers in cars, 2 were viewing the flood waters, 2 were walking, 1 was electrocuted, and 1 died of a heart attack. The activities of the other two victims prior to the flood were unknown.

Many people did not take the rising flood waters seriously. As the water reached the doors of restaurants in Kansas City, the doors were closed. As it broke windows, the occupants retreated to a higher floor. Drivers of automobiles are generally observed to minimize the dangers of driving during flash flood warnings. Many instances were reported in Kansas City, where motorists proceeded through flooded streets when they could see cars stalled ahead. More than 2,000 automobiles were towed after the flood. Thus, it is important to emphasize the importance of proper reactions when driving during flash flood warnings.

WEATHERING

Weathering is a term that indicates how people cope with the weather. Several different weather factors may affect people, including temperature, pressure, humidity, air pollutants and air ionization. According to surveys, about 50% of the population is somewhat sensitive and a smaller percentage is very sensitive to the weather. The geographical location where people live affects their sensitivity. In San Diego, for example, the weather does not change very much and people become acclimated to these small changes. If they move to the midwest, for example, where the weather is more variable, they are more sensitive to weather changes than others who have lived there longer. Sensitivity to weather also changes with age. Babies and people more than 60 years of age are more sensitive than others.

TEMPERATURE

Temperature is one of the primary elements of the atmosphere that affects people. Our bodies try to maintain a constant temperature by such physiological changes as the expansion or contraction of blood vessels. They expand in hot weather to circulate more blood close to the surface of the skin where sweat serves as a cooling mechanism. In very cold weather, the blood vessels contract, and send the blood to deeper parts of the body and this maintains a constant internal body temperature. Our bodies also shiver when we are cold. This is an involuntary response mechanism that helps us to keep warmer.

Heat affects many people emotionally. An increase in violent crimes has been observed in hot weather. Warmer temperatures also tend to dull the intellect. Records were kept for many years on the test scores on civil service exams which showed that more people failed the exam in the months of July and August than any other month.

HUMIDITY

Humidity exerts an effect on people, but it is generally less than the effect of temperature. The combination of high humidity and warm temperature leads to uncomfortable conditions for most people. Less water evaporates from the skin, under these conditions and this is the bodies primary cooling mechanism, since heat from the skin is used in evaporating water.

Humidity also affects the skin directly, since it contracts with low humidity and expands with high humidity. Scar tissue is not as flexible as ordinary skin tissue. When the air is very dry, ordinary skin contracts more than the scar tissue and exerts sufficient pressure to cause severe pain to some people.

Another effect of humidity is exerted through the formation of fog, low visibility and clouds. Some people become depressed under these conditions.

One study showed that when the atmosphere was dry, people checked out more serious books from the library and when it was moist they checked out books for lighter reading. Therefore, people's moods seem to swing with the humidity, cloudiness or their perceived humidity conditions.

PRESSURE

Atmospheric pressure affects parts of the body that are sealed from rapid pressure changes. These include such joints as knees, elbows and shoulders. The brain is also sealed from rapid weather changes and some people have migraine headaches as the pressure changes, while others have aches in their joints.

Severe sunburn, which is more likely to occur with high pressure, causes other effects on a person in addition to the altered skin. More hormones are produced by the pituitary gland and these affect the digestive tract and blood pressure.

A very simple traffic display at a state fair resulted in an interesting conclusion. The exhibit was designed to test a persons reaction time by measuring the length of time, after a light changed from red to green, required for a person to push a button. After a persons reaction time was tested it was displaced for viewing. The number of persons who took the test and their reaction time was recorded. When the results were compared for each day after more than 20,000 people had taken the test it was discovered that the average reaction time was very different each day. Since it was unlikely that the differences could be explained by variations in groups of people each day, other explanations were sought. When the results were compared with the weather on the days when the tests were taken it was found that the fast reaction times corresponded to high pressure conditions. The slowest reactions occurred as a low pressure system was approaching and during the time while it passed. The differences in reaction times were so different that they were valid statistically and no other explanations seemed plausible.

Perhaps the automobile accidents that are more numerous under low pressure conditions that are normally attributed to cloudy skies, lack of visibility, or wet pavement, are also affected even more by a person's reaction time. A person becomes accustomed to a particular automobile and soon learns to stop at an exact distance. If the distance required for stopping under a low pressure system is only one meter greater than with a high pressure system, this may be sufficient to cause numerous accidents. Therefore, it may be worth noting that greater distance can be required for stopping a car when the atmospheric pressure is low, because it may take longer between the time a person perceives an emergency and the time the car actually stop.

PSYCOMETEOROLOGY

A number of symptoms have been identified which are commonly associated with the weather. People who are sensitive to the weather as well as those who are less sensitive report the symptoms. Some of these are tiredness, bad moods, disinclination to work, head pressure, restless sleep, headaches, impaired concentration, difficulty in falling asleep, nervousness, bone fracture pains, and visual flickering.

Psycometeorology deals with the interaction of moods and weather. When people were asked about their general state of health, as they perceived it, they tended to use such terms as "very tired" when the weather was warm and moist and "feel bad" when a change in the weather was coming. They stated that their desire to work declined or their work productiviy was lessened by extremely good weather. Such psycometeorological effects are much harder to substantiate than more direct effects, such as the additional suffering from rheumatoid arthritis during cold and moist weather.

MOOD AND WEATHER

Observations have shown that psychiatric patients were more irritable and in gloomier moods when the skies were cloudy. On the other hand, high pressure systems with sunny skies may produce the opposite effect. One study showed that sunlight suppresses melatonin which is a hormone that tends to produce a depressed state. If sunlight interferes with its production then this represents a direct relationship between the condition of the atmosphere and a person's mood.

Weather changes affect how people relate to each other. Benjamin Franklin, many years ago, advised, "Do business with men when the wind is in the west, when the barometer is high." This remains very good advice, and was verified by a recent study conducted by stopping strangers on the street and asking them a series of questions. The number of questions they were willing to answer before they turned away provided striking results. Americans were much more willing to answer questions on a sunny than on a cloudy day. This difference was attributed to people's moods which were related to the weather.

Some people sink into a more depressed state in the winter months and apparently don't really notice it until spring comes and other people are in a better mood. This is one explanation for the fact that more suicides occur in spring.

Evidence indicates that a person's resistance to viruses is related to mood and mental state and these may be related to weather or other causes. Depressed persons are more susceptible to virus attacks.

RESPONSE TO POSITIVE IONS

Evidence is accumulating that physiological effects can be produced in many people by severe thunderstorms. The positive ion concentration near the ground in a typical thunderstorm is 1,000 to 2,000 ions per cubic centimeter, but varies considerable as lightning discharges occur. Although some people are more sensitive than others, an inhalation of positive ions may increase the level of serotonin in the blood. Serotonin is a powerful hormone that acts in the midbrain on the sleep process, transmission of nerve impulses, and the development of mood.

One investigator, F.G. Sulman, has found that large concentrations of positive ions are associated with hot, dry desert winds. They are also associated with chinook winds.

At least three major effects of high concentrations of positive ions have been identified. The serotonin irritation syndrome occurs when an excess of serotonin is produced by the body. It is characterized by headaches, irritability, sleeplessness, and heart pain.

A second response to large positive ion concentrations is the exhaustion syndrome. In the exhaustion syndrome, the body responds first with an euphoric mood as a result of positive ions triggering the body's release of adrenaline. The adrenaline causes a burst of energy, but this is soon followed by exhaustion. Some unfortunate individuals become unable to function normally as additional weather related symptoms take over.

The third effect is the hyperthyroid response. This occurs when a person's thyroid gland is stimulated and produces excess thyroid hormone. The symptoms are similar to the exhaustion syndrome with an initial burst of energy followed by exhaustion and other reactions such as headache, heart pain or emotional instability.

It is important for people who are extremely weather-sensitive to be aware of the physiological changes that a severe thunderstorm, chinook or desert wind can produce so that their reactions can be anticipated. Counteracting, negative-ion generating machines are available commercially, but it is questionable whether some of them have the capacity to cancel the positive ions in a very large room.

WEATHER SENSITIVITY

Different types of people have been identified in an investigation by the American Bioclimate Institute according to the effect of weather on them. People were grouped according to "cold front people" and "warm front people". The cold front person is typically tall, thin and has long limbs. They find cold air much more unpleasant than other people. The warm front people are generally short, stocky, and have short limbs. They find warm air very uncomfortable.

Most people are not very sensitive to individual variations in responses to weather or to single atmospheric elements such as temperature. Differences in weather sensitivity may affect our interpersonal relationships. Therefore, it is important to know that we don't all react the same way to the weather around us, or even to single atmospheric features such as pressure.

SELF TEST FOR WEATHER SENSITIVITY

A test has been devised to reveal your sensitivity to the weather. Weather sensitivity is determined by summing the numbers assigned to the following questions[†].

Category one is **physique** characteristics.
- If you are lean, slender or lanky, add 3 points. ___
- If you are broad, stocky and stout, add 1 point. ___

Category two is **temperament** characteristics. You gain points from as many questions as apply to you.
- If you tend to me extroverted and jolly, add 1 point. ___
- If you are often emotionally changeable, add 3 points. ___
- If you tend to be easily led or acquiescent, add 3 points. ___
- If you are often irritable or moody, add 1 point. ___
- If you tend to be depressed or pessimistic, add 2 points. ___
- If you are often shy, inhibited or private, add 3 points. ___
- If you tend to be nervous, add 4 points. ___

Category three is **socioeconomic status**.
- If you are professional, or upper class, add 3 points. ___
- If you are blue collar, clerical or laborer, add 3 points. ___

Category four is your **age**.
- If you are 10 to 19 years old, add 3 points. ___
- If you are 20 to 29 years old, add 2 points. ___
- If you are 30–39 years old, add 1 point. ___
- If you are 40–49 years old, add 2 points. ___
- If you are 50–59 years old, add 3 points. ___
- If you are older than 59, add 4 points. ___

Category five is **gender**.
- If you are female, give yourself 3 points. ___

[†] Rosen, S., *Weathering*, 1979, M. Evans & Co., New York.

★ Now add up your **TOTAL POINTS.** ____

Your weather sensitivity is determined by your total score from all five categories. If you scored, 0–5, you are **weather resistant**. Generally you are indifferent to weather changes. If you scored 6–10, you are **weather receptive**; you are aware of your reactions to weather changes. A score of 11–15 indicates that you are **weather sympathetic**; you are never indifferent to weather changes. If you scored 16–20, you are **weather susceptible**; you are always in touch with symptoms that the weather induces. Some people are **weather responsive** (score of 21-25). If you are in this category every weather change is felt in your body. A few people score greater than 25 (**weather keenness**), where severe pain or pleasure accompanies weather changes.

This test was given to several hundred people to determine a typical distribution for the various categories. The results showed that 5% of the people were weather resistant while 34% were weather receptive, and 30% were weather sympathetic. Many people were weather susceptible (22%) and 7% were weather responsive. Only 2% of them were in the weather keenness category as shown in Figure 19-1.

Interviews with individuals who scored in the extreme categories were conducted to provide a check on the results. Those individuals who were rated "weather resistant" paid very little attention to the weather, as the test predicted, while those rated "weather responsive" felt dominated by weather.

CONCLUSIONS

In general, human response to weather varies with the individual and with the type of weather. In both categories the range of voluntary responses is from intelligent, informed responses to unprepared, uninformed responses. It is important to note that voluntary responses can be changed. The person who is caught in a blizzard once is more likely to make better decisions during the threat of the next winter storm. Furthermore, with the proper understanding it is not necessary to experience all aspects of severe and unusual weather in order to respond appropriately.

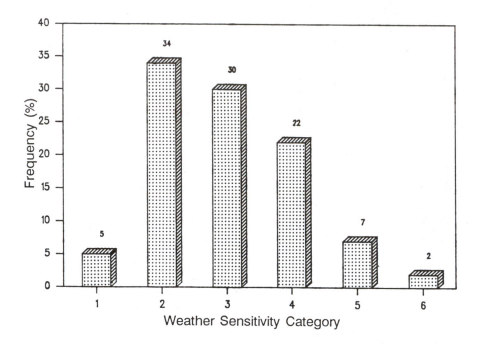

Figure 19-1 The percentage of people who were weather resistant (1), weather receptive (2), weather sympathetic (3), weather susceptible (4), weather responsive (5) or scored in the weather keenness category (6).

Involuntary responses to weather are also very different for individuals and range from weather resistant to weather domination. Active weather factors include positive ions from severe thunderstorms or chinooks, as well as variations of ordinary weather elements.

It may be helpful to understand how weather can affect us in ways beyond our immediate control. Perhaps the information contained in this book will play an important part in helping you make better plans and decisions as you interact with various types of normal, severe and unusual weather in the future.

Additional Reading Suggestions

1. Introduction

Allen, O.E., *Atmosphere*, Time-Life Books, Chicago, Il, 1983.

Canby, T.Y., "Floods - Our Most Precious Resource: Water," *National Geographic*, **153**, Aug. 1980.

Eagleman, J.R., *Meteorology: The Atmosphere in Action*, 2nd ed.,Wadsworth Pub. Co., Belmont, CA, 1985.

Eagleman, J.R., *Weather Concepts and Terminology*, Trimedia Pub. Co., Lenexa, KS, 1989.

Hickox, D.H., "Hot Spots and Cold Spots," *Weatherwise*, **43**:1, 44, 1990.

2. The Largest Storm on Earth

Blackadar, A., "Genesis of Fronts and Air Masses, *Weatherwise*, **38**, 324, 1985.

Eagleman, J.R., *Meteorology: The Atmosphere in Action*, 2nd ed., (Chapter 7, Frontal Cyclones), Wadsworth Pub. Co., Belmont, CA, 1985.

Sassen, K., "The St. Valentine's Day Frontal Passage," *Bulletin of the American Meteorological Society*, **61**:2, 1980.

Schwartz, G.E., "The Day it Snowed in Miami," *Weatherwise*, **30**:2, 1977.

Wheele, D.A, "The Storm on Friday 13 January 1984 in Northeast England," *Weather*, **39**:7, 152, 1984.

More Advanced

Reed, R., "Cyclogenesis in Polar Air Streams," *Monthly Weather Review*, **107**:1, 1979.

Shuman, F.G., "History of Numerical Weather Prediction at the National Meteorological Center," *Weather and Forecasting*, **4**, 286, 1989.

3. Blizzards and Chinooks

Beck, M., Majier, F., and Manning, R., "It Snowed, Snowed, Snowed, (Midwest Blizzards)," *Newsweek*, **93**, Jan. 29, 1979.

Fales, E., Jr., "Whiteout: Winters' Worst Superhighway Nightmare," *Popular Mechanics*, **151**, Jan. 1979.

Gedzelman, S.D., "Mountain Wave Weather in New York City," *Weatherwise*, **30**:11, 1977.

Howard, S.J. "How to Survive a Blizzard When You're on the Road," *Mechanics Illustrated* **80**:90, 1984.

Russell, A., "Frozen Stiff: The Blizzards that Pummel Alberta can be Killers," *International Wildlife*, **12**:14, 1982.

More Advanced

Ellingwood, B., "Probability Models for Annual Extreme Water-Equivalent Ground Snow," *Monthly Weather Review*, **112**: 6, 1170, 1984.

Hardy, R.N., "Estimates of the Probability of Occurrence of High Rates of Snowfall," *Meteorological Magazine*, **104**:1234, 1975.

4. Setting the Stage for Severe Thunderstorms

Fisher, R.E., and Prechtel, A.S., "Fairbanks, Alaska: An Unusual Spring Thunderstorm," *Weatherwise*, **22**, Aug., 1979.

Kessler, E., "Thunderstorms Over Oklahoma," *Weatherwise*, **23**:2, 1970.

Maddox, R.A. and Fritsch, J.M. "A New Understanding of Thunderstorms: The MCC," *Weatherwise*, **37**, 128, 1984.

More Advanced

Carlson, T.N., Anthes, R.A., Schwartz, M., Benjamin, S.G., and Baldwin, D.G., "Analysis and Prediction of Severe Storms Environment," *Bulletin of the American Meteorological Society*, **61**:9, 1980.

Eagleman, J.R., Muirhead, V.U., and Willems, N., *Thunderstorms, Tornadoes, and Building Damage* (Chapter 8), Lexington Books, D.C. Heath and Co., Lexington, Mass., 1975.

Moncrieff, M.W., and Miller, M.J., "The Dynamics and Simulations of Cumulonimbus and Squall Lines," *Royal Meteorological Society Quarterly Journal*, **102**:432, April 1976.

5. Nature of Severe Thunderstorms

Hall, F.F., Neff., W.D, and Frazier, T.V., "Wind Shear Observations in Thunderstorm Density Currents," *Nature*, **264**:5585, 1976.

Serafin, R. and McCarthy, J., "The Microburst, Hazard to Aircraft," *Weatherwise*, **37**, 120, 1984.

Snow, A.D., "Thunderstorms, the Great Destroyer," *Sports Afield*, **183**, April 1980.

Stanford, J.L., "Analysis of a Rare, Westward-Advancing Tornado in Iowa," *Weatherwise*, **27**:5, 1974.

Wood, R., "The Thunderstorm and its Offspring," *Weatherwise*, **38**, 131, 1985.

More Advanced

Eagleman, J.R. and Lin, W.C., "Severe Thunderstorm Internal Structure from Dual Doppler Radar Measurements," *Journal of Applied Meteorology*, **16**:10, 1977.

Klemp, J.B. and Wilhelmson, R.B., "The Simulation of Three-dimensional Convective Storm Dynamics," *Journal of Atmospheric Sciences*, **35**:10, 1978.

Lilly, D.K., "The Structure, Energetics and Propagation of Rotating Convective Storms," *Journal of Atmospheric Sciences*, **43**:113, 1986.

6. The Strongest Storm on Earth

Eagleman, J.R., "Tornado Damage Patterns in Topeka, Kansas, June 8, 1966," *Monthly Weather Review*, **95**:6, 1967.

Fujita, T.T., "Jumbo Tornado Outbreak of 3 April, 1974," *Weatherwise*, **27**:3, 1974.

Galway, J.G. and Schaefer, J.T., "Foul Play–An Undeniably True Tornado Oddity," *Weatherwise*, **32**:3, 1979.

Ganson, R., "Tornado! How Science Tracks Down the Dreaded Twister," *Popular Science*, **203**, Sept. 1973.

Milner, S., "NEXRAD–The Coming Revolution in Radar Storm Detection and Warming," *Weatherwise*, **39**, 72, 1986.

Snow, J.T., "The Tornado," *Scientific American*, **250**, 86, 1984.

More Advanced

Diamond, C.J. and Wilkins, E.M., "Translation Effects on Simulated Tornadoes," *Journal of Atmospheric Sciences*, **41**, 2574, 1984.

Eagleman, J.R, Muirhead, V.U., and Willems, N., *Thunderstorms, Tornadoes, and Building Damage* (Chapter 4, 5, and 9), Lexington Books, D.C. Heath and Co., Lexington, Mass. 1975.

Watkins, D.C., "Electric Discharges Inside Tornadoes," *Science*, **199**, Jan. 13, 1978.

7. Fire from Above

Bailey, B.H., "Ball Lightning Strikes Twice," *Weather*, **39**, 276, 1984.

Bronson, S., "Lightning Protection for Your Home," *Popular Science*, **213**, July, 1978.

Clary, M., "Lightning! Spectacular and Deadly," *Weatherwise*, **38**, 128, 1985.

Krider, E.P., "On Lightning Damage to a Golf Course Green," *Weatherwise*, **30**:3, 1977.

Langa, F.S., "Flashes of Fertilizer in the Summer Sky," *Organic Gardening,* 26, Aug. 1979.

Wood, R.A., "When Lightning Strikes," *Weatherwise*, **41**:4, 206, 1988.

More Advanced

Idone, V.P., Orville, R.E. "Three Unusual Strokes in a Triggered Lightning Flash," *Journal Geophysical Research*, **89**:D5, 7311, 1984.

Johnson, R.L., Janota, D.E. and Hay, J.E., "Operational Comparison of Lightning Warning Systems," *Journal of Climate and Applied Meteorology*, **21**, 703, 1982.

Viemesiter, Peter, E., *The Lightning Book*, M.I.T. Press, Cambridge, Mass., 1972.

8. Ice from the Sky

Changnon, S.A., "Temporal and Spatial Variations in Hail in the Upper Great Plains and Midwest," *Journal of Applied Climatology*, **23**:11, 1531, 1984.

Collins, R.L., "Flying Techniques: Steering Clear of Iceladen Clouds," *Flying*, **100**:3, March 1977.

Curtis, K.D., "Summer's Shrapnel: Hail," *Science Digest*, **74**, July 1973.

Hallet, J., "When Hail Breaks Loose," *Natural History*, **89**, June 1980.

Kennedy, W.S., "Extraordinary Hailstrom in Nebraska: Exerpt from Wonders and Curiosities of the Fairway," *Weatherwise*, **29**, June 1976.

More Advanced

Atlas, D., "Paradox of Hail Suppression," *Science*, **195**, Jan. 14, 1977.

Gokhale, N.R., *Hailstorms Hailstone Growth*, State University of New York Press, Albany, 1975.

Nelson, S., "Influence of Storm Flow Structure on Hail Growth," *Journal of Atmospheric Sciences*, **40**, 1965, 1983.

Ziegler, C.L., Ray, P.S. and Knight, N.C., "Hail Growth in an Oklahoma Multicell Storm," *Journal of Atmospheric Sciences*, **40**, 1768, 1983.

9. The Mighty Middle-Size Storm

Anthes, R.A and Trout, J.W, "Three Dimensional Particle Trajectories in a Model Hurricane," *Weatherwise*, **24**:4, 1971.

DeAngelis, R., "Stalking the Wild Hurricane," *Weatherwise*, **25**:4, 1972.

Frank, N., "When the Next Killer Hurricane Strikes U.S.," *U.S. News and World Report*, **87**, Sept. 17, 1979.

McDougall, G.H., "August 1943: Flying into a Hurricane, *Weatherwise*, **39**, 216, 1986.

Whitton, J., *Disasters, The Anatomy of Environmental Hazards*, University of Georgia Press, Athens, Ga., 1979.

More Advanced

Hasler, A.F. and Morris, J. "Hurricane Structure and Wind Fields from Stereoscopic and Infrared Satellite Observations and Radar Data," *Journal of Climate and Applied Meteorology*, **25**, 709, 1986.

Mogolesko, F.J., "Assessment of Probability of the Probable Maximum Hurricane Event and Its Associated Flooding Potential," *Journal of Applied Meteorology*, **17**, July 1978.

Smith, R.K., "Tropical Cyclone Eye Dynamics," *Journal of the Atmospheric Sciences*, **37**:6, June 1980.

10. Famous Hurricanes and Tropical Storms

Case, B., "Hurricanes: Strong Storms Out of Africa," *Weatherwise*, **43**, 23, 1990.

DeAngelis, R.M., "Enter Camille," *Weatherwise*, **32**:4, 1979.

Hughes, P. "Hurricane Agnes: The Most Costly Storm," *Weatherwise*, **25**:4, 1972.

Thompson, H.J., "The James River Flood of August 1969 in Virigina," *Weatherwise*, **22**:5 , Oct. 1969.

"This Century's Most Tragic Weather Event - The East Pakistan Cyclone," *Weatherwise*, **24**:1, 1971.

Witten, D.E., "Elena," *Weatherwise*, **38**, 259, 1985.

11. Prediction and Modification of Hurricanes

Bradley, D., "Liedtke Syndrome: Decision on Not Seeding the Clouds of Hurricane David," *Motor Boat and Sailing* **144**:8, Dec. 1979.

Falconer, R. and Kadlacek, J., "Hurricane Chemistry," *Weatherwise*, **33**:4, 1980.

"NOAA Spacecraft Tracks Hurricane David's Moves," *Aviation World*, **3**, Sept. 10, 1979.

Pyle, R.L., "Weather Satellite Capabilities: Present and Future," *Weatherwise*, **25**:5, 1972.

Rodenhues, D.R. and Anthes, R.A., "Hurricanes and Tropical Meteorology," *Bulletin of the American Meteorological Society*, **59**:7, July 1978.

More Advanced

Gentry, R.C., "Hurricane Debbie Modification Experiments, August 1969," *Science*, **168**, April 24, 1970.

Jones, R.W., "A Three Dimensional Tropical Cyclone Model with Release of Latent Heat by the Resolvable Scales," *Journal of the Atmospheric Sciences,* **37**:5, 1980.

Neumann, C.J. and J.M. Pelissier, "Models for the Prediction of Tropical Cyclone Motion Over the North Atlantic: An Operational Evaluation," *Monthly Weather Review*, **109**, 522, 1981.

Simpson, R.H. and H. Rieke, *The Hurricane and Its Impact*, Louisiana State University Press, Baton Rouge, 1981.

12. Other Storms and Synoptic Patterns

Bergen, W.R., "Mountainadoes: A Significant Contribution to Mountain Windstorm Damage?," *Weatherwise*, **29**:2, April, 1976.

"Dust Storm Tracked Across U.S.," *Space World*, **11**, Mar. 14, 1977.

Forbes, G.S., "A Storm Chaser Dreamscape," *Weatherwise*, **42**:6, 305, 1989.

Idso, S.B., "Tornado Vortices Spawned by a Desert Brush Fire," *Weather*, **29**:8, 1974.

Idso, S.B., "Whirlwinds, Density Current and Topographic Disturbances: A Meteorological Melange of Intriguing Interactions," *Weatherwise*, **29**:2, April 1976.

Idso, S.B., "Summer Winds Drive Dust Storms Across the Desert; Arizona Haboobs," *Smithsonian*, **5**, Dec. 1974.

Lessard, A.G. "The Santa Ana Wind of Southern California," *Weatherwise*, **41**:2, 100, 1988.

McCollam, J., "Local Winds, Gusts, and Eddies," *Yachting*, **131**, April 1972.

Panofsky, H., "Metamorphic Storms," *Weatherwise*, **33**:3, 1980.

Patric, W. and Pions, N., "Snow Rollers in Minnesota," *Weatherwise*, **24**:5, Oct. 1971.

More Advanced

Bradley, E.F., "An Experimental Study of the Profiles of Wind Speed, Shearing Stress, and Turbulence at the Crest of a Large Hill," *Quarterly Journal of the Royal Meteorological Society,* **106**:447, Jan. 1980.

13. Floods and Drought

Bonnifield, P., *The Dust Bowl: Men, Dirt and Depression*, The University of New Mexico Press, 1979.

Borchert, J.R., "The Dust Bowl of the 1970's," *Annals of the Association of American Geographers*, **61**, 1971.

Eagleman, J.R, *The Visualization of Climate*, D.C. Heath and Co., Lexington, Mass., 1976.

Henry, J.F., "Big Thompson Flood," *Weatherwise*, **29**:6, 1976.

Mogil, M., Monroe, J., and Groper, H., "National Weather Service's Flash Flood Warning and Disaster Preparedness Program," *Bulletin of American Meteorological Society*, **59**:6, 1978.

Namias, J., "The Great Drought of '88," *Weatherwise*, **42**:2, 85, 1989.

Wood, R.A, "Flash Floods," *Weatherwise*, **42**:2, 93, 1989.

More Advanced

Charney, J.G., "Dynamics of Deserts and Drought in the Sahel," *Quarterly Journal of Royal Meteorological Society*, **101**, 1975.

Elsner, J.B., W.H. Drag and J.K. Last, " Synoptic Weather Patterns Associated with the Milwaukee, Wisconsin Flash Flood of 6 August 1986," *Weather and Forecasting*, **4**, 537, 1989.

Hershfield, D., "Secular Trend in Extreme Rainfall?," *Journal of Applied Meteorology*, **18**:8, 1979.

14. Urban and Agricultural Weather Modification

Bach, W., *Man's Impact on Climate*, Elsevier Publishing Co., New York, 1979.

Cleveland, H., "Management of Weather Resources," *Science*, **102**:399, August 4, 1978.

Eagleman, J.R, "A Comparison of Urban Climatic Modifications in Three Cities," *Atmospheric Environment*, **8**, 1974.

Grout, W., "Weather Modification: Cloud Seeding at Ski Resorts," *Ski Magazine*, Dec. 1979.

Plank, V.G.., "Clearing Ground Fog with Helicopters," *Weatherwise*, **22**:3, 1969.

Roberts, W. and Lansford, H., *The Climate Mandate,* Freeman & Co., San Francisco, 1979.

Spar, J. and S. Mayer, "Temperature Trends in New York City," *Weatherwise*, **26**:3, June 1973.

"Windy Cities," *Science Digest*, **79**, Feb. 1976.

More Advanced

Dennis, A.S., *Weather Modification by Cloud Seeding*, Academic Press, New York, 1980.

Hill, G.E., "Dispersion of Airborne Released Silver Iodide in Winter Orographic Clouds," *Journal of Applied Meteorology*, **19**:8, Aug. 1980.

15. Laboratory Tornadoes

Church, C.R., Snow, J.T., and Agee, E.M., "Tornado Vortex Simulation at Purdue University," *Bulletin of the American Meteorological Society*, **58**:9, Sept. 1977.

Peterson, F. and Kesselman, JR., "How Tornado Labs Tame Giant Twisters," *Popular Mechanics*, **154**, Aug. 1980.

Ryan, R.T. and Vonnegut, B., "Miniature Whirlwinds Produced in the Laboratory by High-Voltage Electrical Discharges," *Science*, **168**, June 12, 1970.

More Advanced

Church, C.R., Snow, J.T., "Measurements of Axial Pressures in a Tornado–Like Vortex," *Journal of Atmospheric Sciences*, **42**:6, 576, 1985.

Dessens, J. Jr., "Influence of Ground Roughness on Tornadoes, A Laboratory Simulation," *Journal of Applied Meteorology*, **11**:1, 1972.

Eagleman, J.R., Muirhead, V.U., and Willems, N., *Thunderstorms, Tornadoes, and Building Damage* (Chapters 6 and 7), D.C. Heath & Co., Lexington, Mass. 1975.

Ward, N.B., "The Exploration of Certain Features of Tornado Dynamics Using a Laboratory Model," *Journal of the Atmospheric Sciences*, **29**, 1194, 1972.

16. Softening the Blow

Chapman, F.M., "Tornado House: Shelter Against Storm or Tornado," *Camping Magazine*, **46**:16, Feb. 1974.

Ingersoll, J.H., "House Beautiful Commends Storm Resistant House in New Orleans," *House Beautiful*, **114**, May 1972.

Lee, B., "Extreme Wind Data and Building Design," *Weather*, **33**:11, Nov. 1978.

Melargo, M.G., "Tornado-Safe," *Texas Architect*, **23**:3, 1973.

More Advanced

Eagleman, J.R., Muirhead, V.U., and Willems, N., *Thunderstorms, Tornadoes, and Building Damage* (Chapters 2, 3, 12, 13, 14) D.C. Heath & Co., Lexington, Mass., 1975.

Keller, D., and Lowery, N., "Wind Speeds Able to Drive Straws and Splinters into Wood," *Journal of Applied Meteorology*, **15**:8, 1976.

17. Harvesting the Wind and Sunlight

Duncan, C., "Solar and Wind Power–Some Meteorological Aspects," *Weather*, **32**:12, 1977.

Hewson, E., "Generation of Power from the Wind," *Bulletin of the American Meteorological Society*, **56**:7, 1975.

Kidd, S. and D. Garr, "Can We Harness Pollution-Free Electric Power From Windmills," *Popular Science*, **201**:70, Nov. 1972.

Kocivar, B., "Tornado Turbine Reaps Power from a Whirlwind," *Popular Science*, **210**, Jan. 1977.

Lindsley, E.F., "Wind Power for Home Heating," *Popular Science*, **211**, Nov. 1977.

Rienke, D., "Solar Heating–No Longer a Novelty," *Retirement Living*, **17**, Oct. 1977.

Slinn, W.G.N., "Why Not Cloud Power?," *Bulletin of the American Meteorological Society*, **56**:2, 1975.

More Advanced

McDaniels, D.K., *The Sun, Our Future Energy Source*, John Wiley & Sons, Inc., New York, 1979.

Owens, P., *The Last Chance Energy Book*, John Hopkins University Press, Baltimore, MD, 1979.

18. Acid Rain, Ozone Hole and Greenhouse Effect

Blackadar, A.K., "Smoke Signals, *Weatherwise*, **41**:2, 1988.

Jones, P.D., T.M.L. Wigley, and P.B. Wright, "Global Temperature Variations Between 1861 and 1984," *Nature*, **322**, 430, 1986.

Idso, S.B., "Do Increases in Atmospheric CO_2 have a Cooling Effect on Surface Air Temperatures," *Climatological Bulletin*, **17**, 22, 1983.

Seidel, S., and D. Keyes, *Can We Delay a Greenhouse Warming*, EPA, 1983.

More Advanced

Atkinson, R.J., "Evidence of the Mid-Latitude Impact of Antarctic Ozone Depletion," *Nature*, **340**:6228, 290, 1989.

Henderson-Sellers, A., "Cloud Changes in a Warmer Europe," *Climatic Change*, **8**, 25, 1986.

Hofmann, D.J., J. Harder, S. Rolf and J. Rosen, "Balloon-borne Observations of the Development and Vertical Structure of the Antarctic Ozone Hole," *Nature*, **326**, 59, 1987.

Washington, W.M., and G.A. Meehl, "General Circulation Model CO_2 Sensitivity Experiments: Snow-Sea Ice Albedo Parameterizations and Globally Averaged Surface Air Temperature," *Climatic Change*, **8**, 231, 1986.

19. Human Response to Weather

Rosen, S., *Weathering*, M. Evans & Co., New York, 1979.

Clary, M., "The Human Dimension of Hurricane Forecasting," *Weatherwise*, **40**:4, 196, 1987.

Curry, W., "Camille Revisited. A Critique of Community Response to a Major Disaster," *Hospitals*, **43**:21, 1969.

Hocking, F., "Extreme Environmental Stress and its Significance for Psychopathy," *American Journal of Psychopathy*, **24**, 1970.

Sims, J.H. and Bauman, D.D., "The Tornado Threat: Coping Styles of the North and South," *Science*, **176**, June 30, 1972.

White, G.F., *Natural Hazards: Local, National, Global*, Oxford University Press, New York, 1974.

More Advanced

Krueger, A.P., "Are Air Ions Biologically Significant?" *International Journal of Biometeorology*, **18**, 1974.

Sulman, F.G., "Air Ionometry of Hot Dry Desert Winds and Treatment with Air Ions of Weather Sensitive Subjects," *International Journal of Biometeorology*, 18, 1974.

Index

Aerogenerator, 337, 339
Air mass, 19, 20, 31
Anticyclone, 25
Anticyclonic tornado, 225
Automobiles in tornadoes, 326

Basements, 317
Beneficial competition, 154, 156
Big Thompson Flood, 241, 247, 249, 250
Blizzard, 8, 29, 32-53

Chinook, 50-52, 235
Cloud seeding, 154, 156-159, 284-287
Comfort index, 274, 278
Compression heating, 226
Conservation of Angular Momentum, 10, 11, 31
Coriolis acceleration, 15, 170
Cyclogenesis, 25-27, 31
Cyclone (Bay of Bengal), 179, 200-202

Deaths
    blizzards, 33, 43, 47
    flood, 133
    hailstorms, 137
    hurricanes, 133
    lightning, 133
    tornadoes, 133
Divergence, 25-27
Doppler radar, 77, 81-83
Double vortex thunderstorm, 75-82

Downburst, 233
Drifting snow, 42
Drought, 4, 253-264
    defined, 253
    New York, 254
Dry adiabatic lapse rate, 56
Dry line, 58
Dust devil, 171, 219, 220
Dust storms, 254, 256, 261-264
Dynamic updraft, 79, 93

Easterly waves, 165
Eddy tornado, 225
Electricity
    from wind, 336-342
    in clouds, 124-126
Energy
    sea breeze, 335
    solar, 342-348
    supply, 333
    thunderstorms, 334-336
    wind, 336-342
Evapotranspiration, 259-261

Fatalities
    blizzards, 33
    floods, 133
    hailstorms, 137
    hurricanes, 133
    lightning, 133
    tornadoes, 133

Fertilizer (from lightning), 130
Flash flood warnings, 253, 254
Floods, 3, 6
Fog dispersal, 283
Franklin, Benjamin, 115-117
Freezing rain, 40, 47
Frontal cyclone, 8, 10-31, 62
    March 2-4, 1977, 35
    model, 21
    February 23-24, 1977, 36
    January 26, 1978, 44
    November 27-29, 1978, 39
    June 8, 1974, 84
    January 17, 1979, 23
    March 3, 1979, 23
    structure, 24
Fronts
    cold, 21, 22, 31
    stationary, 45
    warm, 21, 22
Frost protection, 287-289
Frostbite, 48, 49

Glaciation (in clouds), 154, 156
Glaze, 41
Gust front, 74, 234, 235

Haboob, 234, 235
Hail, 137-160
    appearance, 142, 150
    damage, 138
    deaths, 137, 138
    formation, 150, 151
    record size, 143
    seasonality, 141
    suppression, 154-157
    swathes, 154, 155
Hailstools, 151, 152
Hailstorm, 137-160
    cloud heights, 148, 149
    distribution, 140
    green color, 148
    updrafts, 144, 146
    wind shear, 147
Heat waves, 228, 229
Hook echo, 64, 86, 107
Human response
    blizzards, 351, 352
    drought, 354, 355
    floods, 355, 356
        hailstorms, 353
        hurricanes, 354
        involuntary, 356, 357
        lightning, 353
        tornadoes, 352, 353
Hurricane
    Agnes, 183, 189-191, 225, 240, 243, 244
    Betsy, 183-185, 205
    Beulah (1963), 195, 214
    Beulah (1967), 179, 183, 195-196
    Camille, 167, 173, 182, 183, 185-188,
        207, 225
    Carla, 177, 178, 192, 225
    characteristics, 171-176
    Debbie, 167, 214
    dissipation, 176
    Donna, 183, 192-195
    Esther, 213, 214
    formation, 167, 170
    Galveston (1900), 196-200
    Ginger, 214
    names, 182, 183
    origin, 166
    path prediction, 206-209
    paths, 204, 205
    prediction, 203
    safety, 216, 217
    seeding, 212
    stages, 164, 165
    T numbers, 211-212
    winds, 173, 174
Hypothermia, 45

Ice pellets, 40
Ice storms, 45, 47
Inadvertent weather modification, 281
Intertropical convergence zone, 172, 257
Inversion, 40
Inverted barometer, 179, 309
Isobars, 24

Jetstream, 10, 12-19
    cause, 15
    cyclogenesis, 25-27
    location, 15
    patterns, 17, 18, 245, 249, 256, 258
    relation to frontal cyclones, 22, 33-38
    relation to hail, 140
    relation to tornadoes, 68
    ridges, 17, 18, 25, 26, 29, 226-229, 258

Jetstream (continued)
  speed, 16
  trough, 18, 25, 26, 29, 37, 38, 244,
    245

Kansas River flood, 247, 248

Laboratory tornadoes, 6, 7, 292-311
  airflow, 310
  attached to side fan, 306
  from blockage, 299, 301
  from crosswinds, 299-311
  Purdue University, 294, 295
  rope shape, 308
  shells, 307
  traveling, 309
  unconfined, 298-311
Latent heat, 14, 56
Lifted Index, 56-58
Lightning, 3, 114-136
  ball, 121, 122
  bolt from the blue, 119
  channel, 126-128
  deaths, 132, 133
  forked, 117, 118
  ribbon, 119
  rods, 117, 131
  safety, 134, 135
  sheet, 119
  unsafe locations, 135
Longwave cyclone, 33-36
Low level jet, 59, 60

Mesocyclone, 79, 233
Metamorphosis, 224, 225
Midlatitude cyclone (see frontal cyclone)
Mississippi River flood, 241, 244-246
Model houses, 330
Moist adiabatic lapse rate, 56
Moisture tongue, 58, 59
Monomolecular film, 214, 215
Mountainado, 221, 222

National Hurricane Center, 8, 207
National Meteorological Center, 8, 29
National Severe Storms Forecast Center, 8,
    106
Northeastern storm, 38

Ocean swells, 179

Precipitation
  colored, 235-236
  in blizzards, 38, 40
  in frontal cyclones, 28, 29
  intensity, 231, 243
  management, 284-287
  rainfall variability, 259, 262

Rapid City flood, 241, 249, 251-253
Remote sensing (hurricanes), 209, 212
Return streamer, 127, 128
Ridge, 17, 18, 26, 29, 226-229, 258
River basins, 240
Roof damage, 328
Roof orientation, 329
Roof slope, 329

Sahelian drought, 257
St. Elmo's fire, 119
Santa Ana wind, 235, 236
Satellite photographs, 29, 30, 42, 84, 165,
    167, 168, 173, 191, 201, 204, 211,
    263, 344
Severe winters, 229-230
Shelter belts, 289-290
Snow flurries, 42
Snow forecast, 42
Snow squalls, 42
Soil moisture, 259
Solar cells, 343-345
Solar collectors, 344, 345
Solar energy, 342-348
Solar heat
  active, 347
  passive, 347
Solar house, 346
Squall lines, 62, 74, 75
Stability, 56
Stepped leader, 126
Storm spotters, 108
Superbolt, 129

Thermal vortices, 223
Thermoelectric effect, 125
Thunder, 123, 124
Thunderstorm, 13, 73-91
  development, 74
  distribution, 55
  double vortex structure, 75-82, 93-96
  electricity, 124-126

Thunderstorm (continued)
  movement, 82–86
  severe (definition), 74
  splitting, 87, 88
  supercell, 74
Tornado, 3, 5, 7, 8, 12, 13, 92–113
  appearance, 99–101
  damage, 314–325
  damage paths, 96–98
  deaths, 105, 106, 313
  development, 93
  distribution, 101, 105
  forecasts, 8, 71
  in hurricanes, 177
  Lubbock, 323–325
  mesocyclone, 79, 93–95
  noise, 108
  safety, 111, 321, 322
  sky, 110
  Topeka, 314–322
Tropics, 14, 166–170, 257
Trough, 18, 25, 26, 29
Trough cyclone, 37–38

Unusual synoptic patterns, 225–227
Upper air divergence, 67
Upper air inversion, 61
Urban weather, 269–283
  heat island, 89, 270–274
  precipitation, 280
  radiation, 270, 271
  relative humidity, 274
  temperature, 270, 272–274
  wind, 278
  vegetation environment, 274

Vortex, 22, 33
Vortex aerogenerator, 341
Vortex twinning, 219
Vorticity, 27

Warm air advection, 225
Warning
  blizzard, 33, 43
  flash flood, 253, 254
  heavy snow, 42
  hurricane, 217
  tornado, 106–110
Watch
  flash flood, 254
  hurricane, 217
  severe thunderstorm, 69, 70
  tornado, 71
Waterspout, 299, 302
Weak echo vault, 158
Weather
  extremes, 231, 232
  in frontal cyclones, 28–30
  modification, 5–7
  simulation, 5, 292–312
Wind chill, 47, 48
Wind
  machine, 288
  power, 338–342
  profile, 65
  resistant houses, 325–331
  shear, 26, 27, 66, 76
Wing aerogenerator, 340
Winter travel, 49, 50, 52